Springer
Berlin
Heidelberg
New York
Barcelona
Budapest
Hong Kong
London
Milan
Paris
Santa Clara
Singapore
Tokyo

Gennadi V. Glinsky

Failure of Apoptosis and Cancer Metastasis

Springer

Gennadi V. Glinsky, M.D., Ph.D.

Cancer Research Center and Department of Biochemistry
University of Missouri-Columbia *and*
MetaStat, Inc.
Columbia, Missouri, U.S.A.

ISBN: 3-540-540-64560-8 Springer-Verlag Berlin Heidelberg New York

Library of Congress Cataloging-in-Publication data

Glinsky, Gennadi V.
Failure of apoptosis and cancer metastasis / Gennadi V. Glinsky.
p. cm. — (Medical intelligence unit)
Includes bibliographical references and index.
ISBN 1-57059-478-3 (alk. paper)
1. Metastasis. 2. Apoptosis.
3. Cell adhesion. I. Title.
II. Series.
[DNLM: 1. Neoplasm Metastasis--physiopathology. 2. Neoplasm Metastasis--prevention & control. 3. Apoptosis--physiology. 4. Cell Adhesion--physiology. 5. Cell Survival--physiology. QZ 202 G561f 1998]
RC269.5.G56 1998
616.99'407—dc21
DNLM/DLC 98-11142
for Library of Congress CIP

Typesetting: R.G. Landes Company, Georgetown, TX, U.S.A.

SPIN 10681654 31/3111 - 5 4 3 2 1 0 - Printed on acid-free paper

Dedication

To my mom and all cancer victims, with hope that we will be able to reach the target on time...

PREFACE

How can we prevent and cure cancer metastasis? It is my hope that this book will help to find the answer. At least it will describe how we are trying to reach this target, and if you decide that we are on the right path, you are welcome to join us in whatever way you can. It is my trust that one day, hopefully in the not so distant future, we will be able to provide help to those in need.

Two contradictory forces feed metastatic dissemination: (1) the separation of individual cancer cells or small groups of cells from the primary tumor as a consequence of the reduced tendency of tumor cells to stick together; (2) the increased adhesiveness of metastatic cancer, cells that allows them to stick together in the blood of lymph, to form multicellular homotypic and heterotypic aggregates, to interact with end-othelial cells and extravasate. It is clear now that the balance of these two forces could determine the metastatic ability of a given cancer since the adhesive properties of cancer cells may control their sensitivity to programmed cell death. Recent discoveries of the adhesion-dependent regulation of sensitivity to apoptosis (programmed cell death) and superior resistance toward apoptosis of highly metastatic cancer cells[1-3] provide a plausible explanation to a longstanding fundamental problem which is critical to our understanding of cancer biology: why clonal evolution and clonal selection during tumor progression inevitably results in metastasis. The acquisition of certain adhesive properties by metastatic cancer cells would render them more resistant toward apoptosis as well as provide a mechanism that is necessary for extravasation and establishment of metastatic deposits. The current challenge is to further explore the genetic, molecular and cell biological characteristics of metastatic cancer cells that makes them an "instant winner" of the tumor progression marathon.

Is the site specificity of cancer metastasis determined by adhesion mechanisms similar to leukocyte-endothelial cell recognition and adhesion? A striking phenotypic similarity in adhesive behavior of metastatic cancer cells and circulating normal cells was noticed several years ago,[4] and these cells have since been shown to have a very similar molecular anatomy. Recent experiments directly confirm the idea that selectin-mediated adhesion plays an important role in metastatic dissemination. Interactions of E-selectin expressed on endothelial cells with sialyl LeX expressed on metastatic cancer cells provide a molecular basis for tumor cell adhesion to endothelia during colon cancer metastasis. Several lines of experiential evidence support the hypothesis that metastatic colon carcinoma cells may utilize adhesion mechanisms similar to those of circulating normal cells such as neutrophils.

1. It has been documented that the prognosis of colon cancer patients is correlated with levels of sialyl LeX expression on colon carcinoma cells: a high level of sialyl LeX expression was associated with poor prognosis.[5]
2. Experiments utilizing human colon carcinoma cell lines with different metastatic potential demonstrated that highly metastatic cells attached more effectively to the E-selectins expressed on activated human or murine endothelial cells compared to their poorly metastatic counterparts.[6] When poorly metastatic cancer cells expressing low cell surface levels of sialyl LeX were genetically engineered to increase the level of sialyl LeX expression, they adhered more strongly to E-selectins.[7]
3. It has recently been shown that metastatic colon carcinoma cells can adhere to P-selectin under flow conditions.[8]
4. It has been demonstrated that human colon carcinoma cells Colo 205 express a potential selectin ligand sialyl LeA in mucin-type glycoprotein leukosialin.[9]

Thus, one of the key steps in the metastatic cascade—adhesion of circulating cancer cells to endothelia—may involve carbohydrate-lectin interactions constituting the binding of E- and P-selectins expressed on endothelial cells and platelets and the selectin ligands (sialyl LeX and sialyl LeA structures) expressed on metastatic cancer cells. Similar selectin-mediated interactions were shown to be key functional elements in the molecular architecture of leukocyte-endothelial cell adhesion.[10]

Two types of aberrant glycosylation clearly contribute to decreased adhesiveness of cancer cells and their ability to be released from the primary tumor. Modulation of the activity of N-CAM (neural cell adhesion molecule) by polysialic acid is a good example of how weakening cell-cell interaction may allow detachment of cancer cells from primary tumors and increase their migration ability at the initial stage of metastatic dissemination. Polysialic acid is a linear homopolymer of a-2,8-linked sialic acid residues which is mainly attached to the N-CAM. The expression of polysialic acid, a large negatively charged carbohydrate molecule inhibits, and removal of sialic acid increases, N-CAM-mediated homophilic adhesion. Most of the N-CAM molecules in adult tissues lack polysialic acid, in contrast to the polysialated form of N-CAM which is typical for embryonic development.[11-13] Several facts support the idea that polysialic acid modulated, N-CAM-dependent cell adhesion seems to play a significant role in determining metastatic potential of small cell lung carcinoma.[14] N-CAM expression of polysialic acid distinguishes small cell lung carcinoma from carcinoids.[15] In vitro reaggregation ability of two cell variants derived from the NCI-H69 small cell lung carcinoma cell line was correlated with the amount of polysialic acid expression.[14] F3 cells, which express a large level of polysialic acid, exhibit a significantly retarded in vitro reaggregation ability compared to E2 cells, which contain a negligible amount of polysialic acid. This reduced reaggregation ability of F3 cells was restored after enzymatic removal of polysialic acid. Most importantly, in vitro colony forming ability in semisolid media, as well as in vivo metastatic potential in nude mice of polysialic acid-positive clonal variant F3, was markedly higher compared to the polysialic acid-negative cell variant E2.[14] Thus, polysialic acid-dependent weakening of N-CAM-mediated cell adhesion may allow detachment of metastatic cancer cells from primary tumor and may increase their migration potential. The reduced level of N-CAM itself may cause a similar effect. Indeed, the amount of N-CAM is decreased in some metastatic cells.[16,17]

A second molecular cause of the diminished adhesiveness of cancer cells is the most common form of aberrant glycosylation detected in a variety of animal and human cancers. This type of aberrant glycosylation represents an enhanced expression of b1-6GlcNAc antennae in N-linked multi-antennary structures, which results from an increased level of GlcNAc-transferase V.[18,19] Recent experimental evidence strongly indicates that this type of aberration in the glycosylation pattern of the surface of cancer cells may cause a reduction in cell-cell adhesion in tumors, increased release of cancer cells from primary tumors and enhanced migration ability and motility of cancer cells, thus leading to acceleration of metastatic dissemination. It has been demonstrated that reduction in the level of b1-6GlcNAc antennae resulting from bisecting b1-4GlcNAc to the mannose core, induced by GlcNAc-transferase III gene transfection, leads to overall structural changes, enhanced E-cadherin activity, and reduced malignant potential.[20] Increased b1-6GlcNAc antennae resulting from enhanced GlcNAc-transferase V activity decreases cadherin function, diminishes adhesiveness of cancer cells, stimulates release of cancer cells from primary tumors and promotes metastatic dissemination.[21] Disturbance of the E-cadherin/b-catenin function results in modulation of cell-cell and cell-matrix adhesive properties of invasive colon carcinoma cells, leading to decreased cancer cell self-adhesion, increased motility of cancer cells on laminin and collagen substrates and enhanced metastatic potential.[22]

Similar mechanisms may be applicable to modulation of integrin function, since de-N-glycosylation of a5b1 integrin results in dissociation of a5 and b1 subunits and loss of binding to fibronectin.[23] Thus, aggressive metastatic behavior of cancer cells may depend upon enhanced b1-6GlcNAc antennae to form multi-antennary N-linked carbohydrate structures expressed on numerous cell adhesion receptors such as cadherins integrins, CD44, LAMP-1, etc. Aberrant glycosylation of these cell adhesion molecules apparently causes profound alterations in their function, thereby reducing adhesiveness of cancer cells, increasing their release from the primary tumor, enhancing their migration and motility, and promoting their invasive and metastatic abilities. This mechanism could be affected or eliminated by application of N-glycosylation processing inhibitors such as castanospermine, N-methylnojirimycin, and, especially, swainsonine and its analogs. It has been reported that swainsonine reduces a-mannosidase activity, alters N-glycosylation processing and strongly inhibits metastatic ability of MDAY-D2 mouse tumors.[24] The potential cell biological mechanisms of the anti-metastatic activity of swainsonine may include its ability to restore contact inhibition of growth, suppress colony formation in soft agar, and induce normal function of LAMP and integrin cell adhesion receptors.[25] A Phase I clinical trial of swainsonine on patients with advanced cancer was reported to have promising preliminary results.[26] Some lipophilic swainsonine analogs exhibited even stronger anti-metastatic effects.[27]

In recent years the concept of anti-cell adhesion (ACA) therapy of cancer metastasis[4] was supported by significant progress in fundamental and translational cancer research. Particularly dramatic advances have been made in identification and structur-function characterization of cancer-associated carbohydrate antigens such as cell adhesion molecules that play a substantial role in metastatic dissemination of human cancer (reviewed in refs. 28, 29), as well as in the demonstration of a significant anti-metastatic effect against human breast cancer by synthetic glycoamine analogs.[30,31] The concept has emerged from combination therapy of cancer metastasis based on application of cancer cell aggregation and adhesion inhibitors such as antimetastatic synthetic glycoamines and traditional chemo- and radiation therapy. The rationale is to synergize the apoptosis-induction forces of these therapies in order to overcome the adhesion-dependent multicellular drug resistance of cancer cells. Synthetic glycoamine analogs are our leading family of novel anti-metastatic drugs, some of which should enter Phase I clinical trial in 1998. However, the experimental evidence indicates that peptide-based antimetastatic therapeutics and metastasis diagnostic reagents may be only few steps behind. Using a combinatorial chemistry approach, we were able to identify, synthesize, and characterize the specific carbohydrate-binding peptides directed against particular tumor-associated carbohydrate antigens.[32,33] The synthetic peptides are capable of bindong in vitro to human breast and prostate carcinoma cells and inhibiting homotypic aggregation of cancer cells. An intriguing new idea is to develop a cell adhesion modification (CAM) therapy of cancer metastasis based on combined application of N-glycosylation inhibitors, with the goal of restoring normal function of cell adhesion receptors such as integrins and antimetastatic synthetic glycoamines with emphasis on the inhibition of homo- and heterotypic cancer cell aggregation and adhesion mediated by carbohydrate-lectin interactions. CAM therapy of cancer metastasis would target two distinct critical steps in the metastatic cascade and may limit the ability of metastatic cancer cells to find the "escape gate" from the circulatory trap they are destined to enter during metastatic dissemination.

How to prevent and cure cancer metastasis? We do not have the answer yet, but we are seeing light at the end of the tunnel.

References

1. Glinsky GV, Glinsky VV. Apoptosis and metastasis: a superior resistance of metastatic cancer cells to the programmed cell death. Cancer Lett 1996; 101:43-51.
2. Glinsky GV, Glinsky VV, Ivanova AB, Hueser CN. Apoptosis and metastasis: increase apoptosis resistance of metastatic cancer cells is associated with the profound deficiency of apoptosis execution mechanisms. Cancer Lett 1997; 115:185-193.
3. Glinsky GV. Apoptosis in metastatic cancer cells. Crit Rev Oncol/Hemat 1997; 25:175-186.
4. Glinsky GV. Cell adhesion and metastasis: is the site specificity of cancer metastasis determined by leukocyte-endothelial cell recognition and adhesion? Crit Rev Onc/Hemat 1993; 14:229-278.
5. Nakamori S, Kameyama M, Imaoka S, Furukawa H, Ishikawa O, Sasaki Y, Kabuto T, Iwanaga T, Matsushita Y, Irimura T. Increased expression of sialyl Lewis X antigen correlates with poor survival in patients with colorectal carcinoma: clinicopathological and immunohistochemical study. Cancer Res 1993; 53:3632-3637.
6. Sawada R, Tsuboi S, Fukuda M. Differential E-selectin-dependent adhesion efficiency in sublines of a human colon cancer with distinct metastatic potential. J Biol Chem 1994; 269:1425-1431.
7. Sawada R, Lowe JB, Fukuda M. E-selectin-dependent adhesion efficiency of colonic carcinoma cells is increased by genetic manipulation of their cell surface lamp-1 expression levels. J Biol Chem 1993; 268:12675-12681.
8. Goetz DJ, Ding H, Luscinscas FW. A colon carcinoma cell line adheres to P-selectin under flow conditions independent of PSGL-1. Mol Biol Cell 1995; 6:410a.
9. Baeckstrom D, Zhang K, Asker N, Ruetschi U, Ek M, Hansson GC. Expression of leukocyte-associated sialoglycoprotein CD43 by a colon carcinoma cell line. J Biol Chem 1995; 270:13688-13692.
10. Springer TA. Traffic signals for lymphocyte recirculation and leukocyte emigration: the multistep paradigm. Cell 1994; 76:301-314.
11. Edelman GM. Cell adhesion and the molecular processes of morphogenesis. Annu Rev Biochem 1985; 54:135-169.
12. Rutishauser U, Acheson A, Hall A, Mann DM, Sunshine J. The neural cell adhesion molecule (NCAM) as a regulator of cell-cell interactions. Science 1988; 240:53-57.
13. Finne J. Occurence of unique polysialosyl carbohydrate units in glycoproteins of developing brain. J Biol Chem 1982; 257:11966-11970.
14. Scheidegger EP, Lackie PM, Papay J, Roth J. In vitro and in vivo growth of clonal sublines of human small cell lung carcinoma is modulated by polysialic acid of the neural cell adhesion molecule. Lab Invest 1994;

70:95-106.
15. Komminoth P, Roth J, Lackie PM, Bitter-Suermann D, Heitz PU. Polysialic acid of the neural cell adhesion molecule distinguishes small cell lung carcinoma from carcinoids. Am J Pathol 1991; 139:297-304.
16. Linnemann D, Raz A, Bock E. Differential expression of cell adhesion molecules in variants of K1735 melanoma cells differing in metastatic capacity. Int J Cancer 1989; 43:709-712.
17. Andersson AM, Moran N, Gaardsvoll H, Linnemann D, Bjerkvig R, Laerum OD, Bock E. Characterization of NCAM expression and function in BT4C and BT4Cn glioma cells. Int J Cancer 1991; 47:124-129.
18. Yamashita K, Ohkura T, Tachibana Y, Takasaki S, Kobata A. Comparative study of the oligosaccharides released from baby hamster kidney cells and their polyoma transformant by hydrazinolysis. J Biol Chem 1984; 259:10834-10840.
19. Dennis JW, Laferte S. Asn-linked oligosaccharides and the metastatic phenotype. In: Reading CL, Hakomori S, Markus DM, eds. Altered Glycosylation in Tumor Cells. New York: Alan R. Liss.1988:257-267
20. Yoshimura M, Nishikawa A, Ihara Y, Taniguchi S, Taniguchi N. Suppression of lung metastasis of B16 mouse melanoma by N-acetylglucosaminyltransferase III gene transfection. Proc Natl Acad Sci USA 1995; 92:8754-8758.
21. Yoshimura M, Ihara Y, Matsuzawa Y, Taniguchi N. Aberrant glycosylation of E-cadherin enhances cell-cell binding to suppress metastasis. J Biol Chem 1996; 271:13811-13815.
22. Breen E, Steele G, Jr, Mercurio AM. Role of the E-cadherin/a-catenin complex in modulating cell-cell and cell-matrix adhesive properties of invasive colon carcinoma cells. Ann Surg Oncol 1995; 2:378-385.
23. Zheng M, Fang H, Hakomori S. Functional role of N-glycosylation in a5b1 integrin receptor: de-N-glycosylation induces dissociation or altered association of a5 and b1 subunits and concomitant loss of fibronectin binding activity. J Biol Chem 1994; 269:12325-12331.
24. Dennis JW. Effect of swainsonine and polyinosinic:polycytidylic acid on murine tumor cell growth and metastasis. Cancer Res 1986; 46:5131-5136.
25. Demetriou M, Nabi IR, Coppolino M, Dedhar S, Dennis JW. Reduced contact inhibition and substratum adhesion in epithelial cells expressing GlcNAc-transferase V. J Cell Biol 1995; 130:383-392.
26. Goss PE, Baptiste J, Fernandes B, Baker M, Dennis JW. A phase I study of swainsonine in patients with advanced malignancies. Cancer Res 1994; 54:1450-1457.
27. Dennis JW, White SL, Freer AM, Dime D. Carbonoyloxy analogs of the antimetastatic drug swainsonine: activation in tumor cells by esterase. Biochem Pharmacol 1993; 46:1459-1466.
28. Fukuda M. Possible role of tumor-associated carbohydrate antigens. Cancer Res 1996; 56:2237-2244.
29. Hakomori S. Tumor malignancy defined by aberrant glycosylation and sphingo(glyco)lipid metabolism. Cancer Res 1996; 56:5309-5318.
30. Glinsky GV, Price JE, Glinsky VV, Mossine VV, Kiriakova G, Metcalf JB.

Inhibition of human breast cancer metastasis in nude mice by synthetic glycoamines. Cancer Res 1996; 56:5319-5324.
31. Glinsky GV, Mossine VV, Price JE, Bielenberg D, Glinsky VV, Ananthaswamy HN, Feather MS. Inhibition of colony formation in agarose of metastatic human breast carcinoma and melanoma cells by synthetic glycoamine analogs. Clin Exp Metastasis 1996; 14:253-267.
32. Peletskaya EN, Glinsky GV, Deutscher SL, Quinn TP. Identification of peptide sequences that bind the Thomsen-Friedenreich cancer-associated glycoantigen from bacteriophage peptide display libraries. Mol Diversity 1996; 2:13-18.
33. Peletskaya EN, Glinsky VV, Glinsky GV, Deutscher SL, Quinn TP. Characterization of peptides that bind the tumor-associated Thomsen-Friedenreich antigen selected from bacteriophage display libraries. J Mol Biol 1997; 270:374-384.

CONTENTS

CHAPTER 1

Cell Adhesion Cascade as a Multicomponent and Multistep Process with Multiple Decision Forks for Death and Life

1.1. Introduction

Metastasis is one of the most important malignant features of human cancer; it represents the morphological stage of generalization of the disease through the body of the tumor host. Adhesiveness is one of the key cell surface-mediated properties that is altered during malignant transformation,[1] and the abnormal adhesiveness of tumor cells is thought to contribute to the metastatic behavior of these cells. Implicit in the concept of metastasis is the separation of individual cells or small groups of cells from the primary tumor. It has been suggested that the intrinsically low adhesiveness of cancer cells contributes to separation.[2] Coman and others[3-5] have observed that tumor cells were more easily separated from solid tumors than were normal cells from corresponding tissues. Tumor cells have also been shown to be less adherent than normal cells to artificial substrates.[6,7] Therefore, the reduced tendency of tumor cells to stick together is important for metastatic dissemination.

On the other hand, homotypic and heterotypic aggregation properties of tumor cells represent important biological features of malignancy, because these properties of transformed cells could determine the metastatic ability of neoplastic tumors.[8,178,248-253,256,257,265-269] The key structural determinants of the tumor cells that participate in cell recognition, association and aggregation have been determined as carbohydrates and carbohydrate-binding proteins.[178,248,249,264-270,279] The concentration and size dis-

Failure of Apoptosis and Cancer Metastasis, by Gennadi V. Glinsky.

tribution of tumor cell clumps that enter the circulation play a significant role in the metastatic process.[8-11] Intravenous injected tumor cells in clump form have a greater tendency to form metastases than do the same number of single tumor cells.[8,11,12,251] Therefore, the ability of cancer cells to display certain adhesive properties is also important for metastatic spreading.

It has been suggested that the key features of cancer cell adhesion are: (a) preservation of cell recognition function and the initial reversible steps of cell-cell or cell substrate adhesion; and (b) impairment of the ability to display secondary stable attachment, strong adhesion, and terminal tissue specific cell-cell and cell-substrate contacts.[13,164] The profound defects in protein adhesive systems, primarily mediated by the cadherin[14-17] and integrin[18-21] families of adhesion receptors is characteristic of malignant transformation and may contribute significantly to the abnormal locomotion, motility, invasion and metastasis of cancer cells. However, the acquisition of certain adhesive properties by malignant cells is extremely important for invasion, motility and metastasis.[12,13,21] Evidence will be considered that the metastatic cancer cells typically lose the adhesive characteristics of their parent coherent tissue, but acquire the certain adhesive properties of embryonal and/or circulating cells. This subpopulation of cancer cells may play a key role in determination of the target organ(s) for distant colonization and secondary tumor(s) formation.

The cell-biological and molecular mechanism of site specificity of cancer metastasis remains a mystery. The traditional lock-and-key model predicts that cancer cells may have a specific membrane-associated structure(s) which determines specificity of cancer cell-endothelium recognition and site(s) of metastatic deposition. Here, we will summarize the evidence for an alternative concept: metastases do not occur as a result of random survival of cancer cells released from the primary tumor but from selective growth of specialized subpopulations of highly metastatic cells with certain specific properties.[326] However, these "primary" or "key" metastatic cells would need support from the "secondary" or "helper" cells, both nonmetastatic cancer cells and normal cells, in order to complete each step of the metastatic cascade. Evidence will be discussed for whether similar carbohydrate cell adhesion determinants on the surface of both cancer and host cells may be responsible for either a "seed and soil" mechanism of metastasis or for spreading of cancer cells on the basis of circulatory anatomy. According to this model, the site specificity of cancer metastasis is determined by mechanisms similar to leukocyte-endothelial cell recognition and adhesion. Homotypic and heterotypic cancer cell aggregation in blood vessels and subsequent formation of "multicellular metastatic units" will be considered as prerequisite to cancer metastasis. This "multicellular metastatic unit" consists of tumor cells (both capable and incapable of metastasizing individually), leukocytes, and platelets, where site specificity of endothelium recognition, adhesion and stable attachment could be determined by a subset(s) of leukocytes involved in the formation of the "multicellular metastatic unit". This particular subset of leukocytes may serve as "carrier cells" targeting the metastatic deposit(s) to the specific site(s) of secondary tumor foci formation.

The prospect of anti-cell adhesion (ACA) therapy of cancer metastasis will be discussed, with emphasis on the inhibition of carbohydrate-mediated homo- and heterotypic cancer cell aggregation. Evidence will be presented that the search for the carbohydrate-based homotypic and heterotypic cancer cell aggregation inhibitors was successfully completed at the first stage of a new antimetastatic drug design. Inhibition of extravasation of cancer cells, and blocking of their homotypic and heterotypic adhesion, prevented escape of metastatic cells from blood into the tissues and caused a significant reduction of metastasis. Targeting the metastatic cancer cells in the circulation should be a very effective approach to control metastatic disease, because the number of tumor cells in the circulation declined very rapidly after IV injection: after 24 h, 90-99% had disappeared[22-28] and after 3 days generally less than 1% remained.[23,26,27] Only fewer than one in 10,000 of the tumor cells survive in the circulation and within 8 to 24 hours invade the vessel wall and extravasate.[29] The vast majority of the cancer cells die in the capillaries or lymph nodes mainly victims of mechanical turbulence in the circulation as well as activated apoptosis. Four prospective approaches have been considered for anti-cell adhesion therapy of cancer metastasis:[164] (a) dvelopment of carbohydrate-based competitive cell adhesion inhibitor(s); (b) biospecific inhibitory monoclonal antibody and/or lectin-based anti-cell adhesion therapy; (c) search for the factor(s) of down-regulation of corresponding carbohydrate and/or carbohydrate-binding (lectins) structures on the surface of cells; (d) gene-therapeutic approaches based on activation/inactivation of expression of gene(s) coding corresponding adhesion receptor-counterreceptor pairs, glycosyltransferases, mediators of adhesive intracellular signaling network and receptors crosstalk, e.g., protein phosphorylation pathways, components of cytoskeletons etc.

1.2. Cell Adhesion as a Cascade Process with Cooperative and Combinatorial Adhesive Specificity

For almost a century the lock-and-key hypothesis has been one of the central dogmas of biology. Developing from enzymology (Emil Fischer), immunology (Paul Ehrlich), and embryology (Frank Rattray Lillie), this concept has become a dominant idea in biochemistry and cellular biology, including an explanation of cell-cell recognition phenomena. During recent years dramatic progress has been made in our understanding of molecular mechanisms of cell adhesion, particularly in identification and structural-functional characterization of several cellular receptor systems and their ligands participating in a multicomponent and multistep "cascade-like" cell adhesion reaction. Therefore, considerable revision of a simple lock-and-key model of cellular adhesion was required. Cellular adhesion consists of multiple steps; it is a multicomponent "cascade-like" process that requires sequential activation and inactivation of a number of different receptor-counterreceptor pairs (Fig.1.1) and involves cell adhesion receptor cross-talk, as well as intracellular signaling and intercellular communication among the interacting cells.[13,20,30,31] The specificity of most well studied adhesive events arises not from high selectivity of individual adhesion receptors but from a unique sequential combination of several different receptors acting cooperatively.[20,30,31] This cooperative (e.g., different adhesion receptors being expressed and acting cooperatively on the same stage of cellular adhesion) and combinatorial (e.g., unique sequential combination of activation and inactivation of different adhesion receptors being expressed and acting on the different steps of adhesive cascade) adhesive specificity requires involvement of several not highly specific individual adhesive receptors to give highly selective cell adhesion during development, physiological processes (lymphocyte traffic, wound healing, hemostasis), and pathological conditions (metastasis, inflammation, thrombosis).

Interacting cells coexpress on their surface the multiple families of different molecules involved in cell-cell type specific recognition, association, and adhesion. The function of multiple cell adhesion molecules on a single cell during specific cell-cell adhesion has been explored in several experimental systems.[32-35] It has been shown, for example, that the extension of neurites by peripheral motor neurones on the surface of myotubes involves multiple mechanisms for growth and guidance acting in concert during development.[33] This process involves neuronal extracellular matrix receptors (particularly, for laminin), as well as Ca^{2+}-dependent (neural Ca^{2+}-dependent cell adhesion molecules) and Ca^{2+} independent (neural cell adhesion molecules) cell adhesion molecules on neurite surfaces. The neuronal receptors for extracellular matrix proteins and different cell adhesion molecules, used by neuronal growth cones for interaction with myotubes, each provides a distinct but acting in concert mechanism for regulation of growth cone motility.[33] One constituent of such a multicomponent cell adhesion system is the cell membrane carbohydrates. Depending on the structure of the carbohydrates, they may act either as inhibitors or activator of cell-cell recognition and adhesion.

The neural cell adhesion molecule (NCAM), one of the most abundant adhesion molecules, can influence a wide variety of diverse cellular events: junctional communication, axon association with pathways and targets, and signals that alter levels of neurotransmitter enzymes.[32] These pleiotropic effects reflect the ability of NCAM to regulate membrane-membrane contact of interacting cells, required for initiation of specific interactions between other molecules. Such regulation can occur through changes in either NCAM expression or the content of polysialic acid (PSA) on NCAM molecules.[32] When NCAM with a low PSA content is expressed, adhesion is increased, the probability of junction formation from interacting subunits of other adhesion molecules is enhanced by increasing the extent or duration of membrane-membrane contact, and contact-dependent events are triggered. The large excluded volume of NCAM PSA may inhibit cell-cell interaction through hindrance of overall membrane apposition,[32] and the initiation of cellular interactions via specific ligand-receptor pairs may occur as a result of reduction in NCAM itself and/or NCAM PSA content. The probability of receptor-ligand interactions is enhanced by reducing the excluded volume of carbohydrate between membranes, which otherwise impedes close cell-cell contact. These steric alterations of the overall degree of close membrane-membrane contact could allow differential regulation of the function of several ligand-receptor

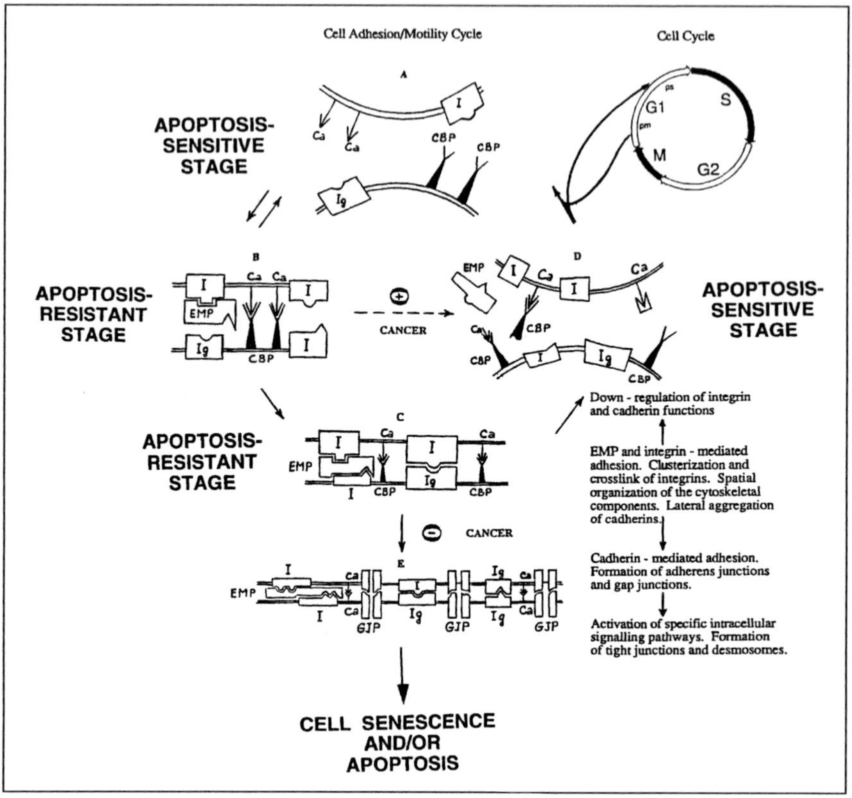

Fig. 1.1. A model of cell adhesion-cell motility and/or cell multiplication cycle discussed in the text. Stages A,B,C and D comprise a reversible cell adhesion/motility/multiplication cycle.
Stages A,B,C, and E comprise an irreversible (?) cell adhesion/stable attachment/terminal adhesion cycle.
Stages A and D represent preadhesion or competent state and postadhesion or detachment state of interacting cells, respectively. Stages B,C and E indicate the initial reversible adhesion, secondary stable attachment, and tissue-specific terminal (irreversible?) adhesion, respectively.
Stage B is primarily mediated by carbohydrate (Ca)- carbohydrate binding protein (CBP) interactions with the involvement of carbohydrate-carbohydrate and integrin (I)- extracellular matrix protein (EMP) interactions in some cellular systems. Initial weak homophilic cadherin-cadherin interactions may also occur. Stage C is primarily mediated by protein adhesion systems comprised of integrin (I), immunoglobulin (Ig), extracellular matrix protein (EMP) and cadherin families of adhesion molecules. At stage E the tissue-specific structural proteins become involved and gap junctional intercellular communication channels are opened (GJP-gap junction proteins). Cancer cells lose the ability to display stage E of cell type-specific adhesion. Stage C, which is largely dependent upon integrin-mediated functions, would give anchorage dependence of growth for normal cells. Subversion or replacement of the need for extracellular matrix protein generated signals in cancer cells may lead to the anchorage independence of growth, and neoplastic cells would pass from stage B to stage D of the cell adhesion/motility/multiplication cycle. At stage D (postadhesion or detachment state) multiple anti-adhesion mechanisms are displayed: (A) dramatic conformational and affinity

systems according to their properties, such as size, mobility and quantity on the cell surface, etc.[32] Most recent experimental evidence clearly indicated the significance of NCAM-mediated adhesion in lung cancer metastasis (reviewed in ref. 165).

The process of cell-cell recognition, association and aggregation is not only multicomponent, but also consists of multiple steps, and a number of models of such a multistep process have been proposed.[13,20,30,31,36] Generally, the initial step in the process is specific recognition between two cells, in which multivalent homo- and heterotypic carbohydrate-mediated interactions play a major role.[13,30,36,37] Initial cell recognition through carbohydrate-carbohydrate or carbohydrate-protein (selectin) interaction is followed by secondary adhesion. This secondary adhesion is primarily mediated by Ca^{2+}-sensitive adhesion molecules or sugar-binding proteins, or by proteins of the immunoglobulin superfamily, or by pericellular adhesive meshwork proteins consisting of fibronectin, laminin, and their receptor systems (integrin). The third step of cell adhesion is the establishment of intercellular junctions, e.g., "gap junctions," in which a cell-cell communication channel is opened through a specific structural protein.

The "sorting-out" behavior of homotypic cell populations plays a major role in morphogenesis and organogenesis, and in this context cell adhesion is processed mainly through recognition between similar types of cells. Specific interactions between identical or simi-

Fig. 1.1. Cont.
changes in components of protein adhesion systems (for example, phosphorylation of integrins may cause apparent inactivation of the receptors); (B) shedding and/or structural modification of glycodeterminants of cell adhesion as well as shedding of carbohydrate binding proteins (for example, shedding of selectins occurs at the second stage of leukocyte-endothelial cell adhesion); (C) free Ca and CBP would block their cellular counterparts and inhibit the cell-cell or cell-substratum adhesion.

Stage D is followed by a cell motility and/or cell multiplication stage that may restore the adhesion competence of cells (stage A). Some cell types may need to enter into the cell multiplication cycle in order to restore the competent adhesive mechanisms. At stage C, the clasterization and cross-linking of integrins brings them together in the plane of the membrane in order to organize specific cytoskeletal structures. The organized cytoskeletal components would support the lateral interactions of cadherins, anchoring them to the cell-cell contact zone. The lateral aggregation of cadherins at the cell-cell contact sites will increase the strength of cellular adhesion, enhance cell-cell binding forces and will lead to the extensive expression of junctional communication, particularly adherins junctions and gap junctions. Subsequently, the development of tight junctions and desmosomes would complete the cellular adhesion. Adapted from reference 164 with permission.

lar carbohydrate chains which are highly expressed in the same types of cells may provide a better explanation for sorting out than the interaction of relatively nonspecific adhesive molecules[36,37] (reviewed in ref. 166). It is important to note that tumor cells retain, to a great extent, their ability to reveal initial stages of cell adhesion; at the same, time cancer is characterized by profound disturbances in the subsequent stages of cell adhesion, with of involvement of extracellular matrix proteins, and followed by the formation of specific "gap junctions." We suggested[13,38-42,164] that recognition between identical types of cells, which provides the basis for sorting a single homotypic population from a heterotypic assemblage of cells, plays a major role in the metastatic spreading of tumor cells. This mechanism provides a basis for the formation of tumor cell multicellular aggregates in the circulatory channels of the tumor host, and for the subsequent arrest of these multicellular aggregates in blood vessel of target organs. Carbohydrate-mediated cell-cell type specific recognition[36,37,43-48] and/or cadherin-mediated homophilic adhesion[31,49-52] both have been suggested as key mechanisms of homotypic sorting out of a cell population from heterotypic assemblage of cells during development. Cadherin mediated adhesion is greatly diminished in many types of cancer cells[14-17] and this mechanism has been proposed as a molecular pathogenetic factor in cancer invasion and metastasis.[51] Therefore, a carbohydrate-mediated adhesion system is the primary candidate for facilitator of the homotypic and heterotypic aggregation of cancer cells.

Certain carbohydrate determinants of the blood group antigen (BGA)-related glycoantigens are the structural determinants participating in the homotypic aggregation of histogenetically different types of cells.[13,42,53,54,164-166] BGA-related glycoepitopes are directly involved in the homotypic (tumor cells, embryonal cells) and heterotypic (tumor cells-normal cells) formation of cellular aggregates (e.g., Lewis X antigens; H-antigens, polylactosamine sequences; and T- and Tn-antigens), which has been demonstrated in different experimental systems.[43-46,55,56] Lymphoid cells (lymphocytes, granulocytes, monocytes) are Lewis family glycoepitope-positive and contain T-antigen (demasked after treatment with neuraminidase).[57-59] Neutrophils bear LeX both on glycolipids and at the termini of N- and O-linked oligosaccharides.[60,61] Carbohydrate ligands for selectins—main adhesion proteins of circulating cells (e.g., leukocytes, platelets) and endothelium—have been described: sialosyl LeX, sialosyl LeA, and LeX have been identified as recognition structures for selectins.[62-65] The selectin ELAM-1 is involved in the differential attachment to endothelium, and subsequent migration into the tissues of specific lymphocyte subsets ($CD4^+$ memory T cells).[66,67]

Thus, recent experimental evidence has generally supported the concept that some of the BGA-related glycodeterminants, which were identified earlier as tumor-associated carbohydrate antigens (TACA), function as key adhesion molecules.[13,42,53,54,164-166] These studies have shown that cell adhesion, through carbohydrate-carbohydrate or carbohydrate-selectin interactions, occurs at the early initial stages of the "cascade" multistep cell adhesion mechanism, and this reaction is a prerequisite for subsequent steps of cell adhesion directed at integrin—or immunoglobulin—based adhesion.[48,68,69,167] Usually, cells coexpress on their surface the multiple components involved in the "cascade" cell adhesion mechanism, and thus this multistep adhesion reaction could be triggered by initial carbohydrate-carbohydrate or carbohydrate-selectin interactions. Such conclusions have been made on the basis of studies of B16 melanoma cell adhesion[48,54] and investigations of leukocyte roll on a selectin.[68,69] Evidence has been presented that specific glycosphingolipid-glycosphingolipid interaction initiates cell-cell adhesion, and may cooperate synergistically with other cell adhesion systems such as those involving integrins[47,48,53,54] (reviewed in ref. 166).

1.2.1. A General Model of Cellular Adhesion

Thus, specific membrane-associated carbohydrate determinants, particularly BGA-related glycoepitopes, function as ligands and/or receptors in cell-cell recognition, association, aggregation, and adhesion. Initial carbohydrate-mediated cell-cell recognition and interaction may trigger the subsequent steps of the "cascade" cell adhesion mechanism and involve the multiple family of cell adhesion molecules in the formation of the stable and specific cell-cell contact. It has been suggested that specific glycosphingolipid-glycosphingolipid interactions may be the general mechanism of the initiation of cell-cell adhesion.[47,48,53,54] This model may be correct for some specific cell types, like melanoma cells, that contain an unusually large amount of gangliosides on the cell surface. However, in the most cases the specific cell-cell recognition has to be initiated by transmembrane glycoproteins with relatively large extracellular carbohydrate-carrying and/or-binding domains.

On a molecular scale, a model of the distances over which adhesion receptors would mediate cellular interactions has been constructed.[70] This model allows prediction of how close together two cell membranes, or a cell membrane and the extracellular matrix, would have to come to allow effective interactions. There are two classes of adhesion interactions that differ significantly in the distance between the plasma membrane of the two closely apposed cells.[70] The first type of cell-cell adhesion contacts is predicted to occur within an intermembrane distance roughly equal to 30 nm, and the second class of intercellular adhesion interactions is predicted to occur within an intermembrane distance of 10-15 nm, which is in good agreement with electronic microscopic measurements of intermembrane distance of membrane apposition in interactions of mitogen-stimulated T cells with target cells.[71] For example, migrating cells do not posses focal contacts, which have a cell to substrate distance of 10-15 nm, but have close contacts with a cell to substrate distance of 30 nm,[72] in good agreement with the distance predicted for FLA-1/ICAM interaction, and for integrin-matrix interaction.[70] Furthermore, the scale of glycocalyx has to be considered, which will oppose cellular interaction due to negative charge repulsion and the loss of entropy involved in compressing or interdigitating the glycocalyces of two contacting cells.[73] The sizes of two major components of the leukocyte glycocalyx, which bear most of the cell surface sialic acid,[74] are about 30 nm.[70] On the time scale, it is more likely that the first type of cell-cell adhesion contacts will occur at the early initial stage of cell-cell recognition and adhesion, and this step more likely will be mediated by transmembrane glycoproteins with relatively big extracellular domains. The concept of the crypticity of glycolipid antigens supports this suggestion.

According to the minimum energy conformation model, the axis of the carbohydrate is perpendicular to that of ceramide; glycolipids are inserted in the lipid bilayer through their ceramide moiety, and their carbohydrates are laid on and fixed to the surface of the lipid bilayer, exposing the hydrophilic surface toward the outside of the membrane.[75,76] According to this model, crypticity of glycolipid antigens will be determined and controlled by several complex factors, such as membrane proteins and sialosyl glycoconjugates surrounding glycolipids, ceramid composition, etc.[77] Therefore, glycolipids are more likely to be involved on the latest stages of cell-cell adhesion when cell-cell contact distance between the plasma membranes of the two closely apposed cells is 10 nm or less. One of the important questions is: what is the regulatory mechanism responsible for sequential cooperative action of the different cellular adhesion systems during cell-cell recognition, association, and stable attachment?

Considering the molecular scale and time scale models, the molecular topography of primary adhesion molecules as well as their rapid down-regulation after initiation of cell-cell contact appear to be critically important. For example, neutrophil extravasation at sites of acute inflammation may be considered as a two-step process.[69,78] First, neutrophils within the flowing blood must recognized and bind venular endothelial cells activated by a tissue-based inflammatory stimulus. The second step involves adhesion strengthening, ensuring the complete stop of the rolling neutrophils, followed by transmigration through the vascular wall. The leukocyte selectin (LECAM-1), and the vascular selectins (ELAM-1 and GMP-140) are primarily responsible for the initial recognition/binding interaction of neutrophils and endothelial cells.[68,69,79,80] The β2 integrin adhesion pathway is primarily involved in the second step of this process and appears to be critically involved in the adhesion strengthening and transmigration stages of the extravasation process.[79,81-84] The expression of LECAM-1 on neutrophils is extremely sensitive to neutrophil activation by cytokines, chemoattractants, and other inflammatory agents:[85-87] complete shedding of LECAM-1 occurs within 5 min of exposure of murine neutrophils to recombinant C5a at 37°C.[85] The same factors that cause LECAM-1 shedding also cause upregulation of β2 integrin function.[82,83] After initial interaction of neutrophils with endothelial cells, largely selectin mediated, LECAM-1 shedding would occur, downregulating a major portion of the selectin-mediated adhesion to facilitate or trigger the integrin-mediated stable attachment and subsequent diapedesis.[78] Down-regulation of selectin function by LECAM-1 shedding may also serve as an autoregulatory mechanism limiting or blocking excessive neutrophil-endo-thelial cell interaction after severe inflammatory stimuli. The critical importance of the molecular topography of adhesion molecules for initiation of cell-cell contact has been shown by Picker et al.[78] The neutrophil selectin LECAM-1 participates in the earliest interactions of neutrophils with endothelial cells through presentation of oligosaccharide ligands to the inducible vascular selectins (ELAM-1 and GMP-140). The enhanced function of LECAM-

1 associated oligosaccharides (sialosyl LeX) reflects the striking concentration of LECAM-1 on neutrophil surface microvilli, the site of initial cellular contact.[78] One of the mechanisms of the cooperative action of different families of adhesion molecules is the processes designated as mutual capping and/or mutual cocapping.[88,89]

The initial binding of a cell adhesion molecule to its receptor under appropriate conditions may result in dramatic redistributions of integral membrane molecules within the fluid membranes of the two contacting cells.[90] The concentrations of the ligand and receptor molecules should be sufficiently large, and the intrinsic rate constant for the dissociation of the ligand-receptor bond should be sufficiently small, to allow the formation of a small number of initial ligand-receptor bonds which can maintain a localized cell-cell contact long enough for the diffusion of many more ligand and receptor molecules of the same specificity (mutual capping) or a variety of different classes of cell adhesion molecules that contact cells may simultaneously exhibit (mutual cocapping). These ligand and receptor molecules, after diffusion into the contact region, will form additional bonds clustered within an extended contact area. The function of mutual capping and cocapping is the increase in concentration of corresponding cell adhesion molecules precisely within the morphologically defined region of cell-cell contact.[89] This clustering phenomenon occurs on because the formation of the initial cell-cell contact as a result of interaction of first ligand-receptor pairs would substantially decrease the free energy of formation of later ligand-receptor bonds. Capping phenomena and closely related processes have been observed experimentally,[91,92] and described theoretically.[73] It has been suggested that the mutual capping and cocapping always occur during cell adhesion, unless molecular mobility in the membrane is inhibited.[89] These capping processes may not only play an important functional role in adhesion phenomena, but also can provide a mechanism to allow the stable binding of certain weak ligand-receptor pairs (for example, specific glycosphingolipid-glycosphingolipid interactions or other carbohydrate-carbohydrate recognition and association) that are thereby allowed to participate in intracellular signaling between two contacting cells. In this context it is interesting to note that the role of a certain sphingolipid catabolite (sphingosine) as a negative modulator of protein kinase C was proposed.[93,94]

A recent study suggests also that sphingosine enhances production of phosphatidic acid, which may trigger intracellular Ca^{2+} mobilization and activate G-protein.[95] The involvement of endogenous ganglioside GM3 in inhibition of epidermal growth factor receptor kinase activity was demonstrated by Weis and Davis.[96] The studies of various glycosphingolipid catabolites show that they may act as an inhibitor and/or activator of protein kinase C and epidermal growth factor receptor kinase.[97,98] Finally, a large body of evidence has established the role of certain phospholipids, such as diacylglycerol and inositol tri- or diphosphate, as a key component of mitogenic signaling pathways that act by modulating protein kinase C activity and Ca^{2+} mobilization.[99-101] The observations that some glycosphingolipids and their catabolites affect transmembrane signaling also support the idea that specific glycosphingolipid-glycosphingolipid interactions may occur in the latest stages of cellular adhesion, in order to participate in modification of intracellular signaling pathways of contacting cells. At least five major plasma membrane glycosphingolipids or phospholipid components, nine identified catabolites or derivatives, and a large number of presently unknown compounds may be involved in the control of transmembrane signaling[54] that provides an unique array of "messenger" molecules for highly specific modulation of intracellular signaling and, possibly changes in gene function following cell-cell type specific adhesion.

The concept of competition between nonspecific repulsion and specific bonding during cell adhesion implies the existence of a repulsive barrier preventing adhesion that must be overcome by bridging molecules, as well as the existence of two phase transitions corresponding, respectively, to the onset of stable cell adhesion and to the onset of maximum cell-cell or cell-substrate contact.[73] A certain configuration of cell-cell contact involving redistribution of receptors into the contact region might be required to achieve a degree of bridging sufficient to overcome the repulsive barrier, and adhesion receptors will become concentrated in regions of cell-cell or cell-substrate contact.

The capping phenomena may also play an important role in the initial stage of cell adhesion, as a mechanism which allows exclusion of the glycocalix components from the site of cell-cell contact. For example, the initial cell-cell contact would be mediated by circulating naturally occurring anticarbohydrate antibodies in order to overcome the size limitation by glycocalyx molecules on the interaction of cell adhesion molecules.[164] Then the mutual capping and cocapping processes will allow initiation of the subsequent stages of cellular adhesion with the involvement of selectin-carbohydrate interactions, immunoglobulin-, integrin-, and extracellular matrix protein-mediated attachment, carbohydrate-carbohydrate recognition and interactions, gap junction formation, and activation of transmembrane and intracellular signaling mechanisms. The experimental evidence presented above clearly shows that BGA-related glycodeterminants may serve as ligand and/or receptor in cell-cell recognition and adhesion. An important unanswered question is whether or not membrane-associated carbohydrates may function as a fully competent cellular receptor and mediate the internalization and cell uptake of corresponding ligand as well as activate a specific intracellular signal transduction pathway.

1.2.2. The Role of Integrins in Cell Adhesion

The earlier definition of integrins as fibronectin or vitronectin receptors has proven too restrictive, because there are numerous receptors for each of these glycoproteins, and they are not highly specific for individual adhesive ligands.[20] There are now about 20 integrins—the major family of receptor molecules by which cells attach to extracellular matrices and/or to other cells. Most cells express several integrins, and most integrins are expressed on a wide variety of cells. The studies of ligand and adhesive specificity of individual integrins using cell adhesion assay, monoclonal antibodies, and affinity chromatography revealed that individual ligands are recognized by more than one integrin, and individual integrins can bind to more than one ligand.[20] The molecular transmembrane topography of integrins indicates that the N-terminal domains of a and b subunits combine to form a ligand-binding head of each integrin. Two cytoplasmic domains interact with cytoskeletal proteins and perhaps with other cytoplasmic components. There is some evidence that different integrin subunits may mediate differing cellular responses, even to common extracellular adhesion ligands.[102-104] Most integrins interact with the actin-based cytoskeleton, but some of them more likely interact with intermediate filaments, which are associated with hemidesmosomes.[105-107]

Individual cells can vary their adhesive properties by selective expression of integrins and by modulation of the binding properties (affinity and ligand specificity) of integrins. For example, $\alpha_2\beta_1$ integrin on many cells can recognize both collagen and laminin,[108,109] but on platelets it is specific for collagen and not laminin.[110] The specificity and affinity of a given integrin receptor on a given cell are not always constant, and numerous examples of modulation of integrin functions include both activation and deactivation of integrins.[20] Often, only after cell activation by specific factors do the integrins become effective receptors for corresponding ligands. The effective activation stimuli for β_2 integrins that are expressed on leukocytes vary depending on cell type. Activation can be accomplished by phorbol esters,[111] by adherence to fibronectin,[112,113] or by various inflammatory mediators and cytokines (see below). There is an adhesive cascade leading to attachment of neutrophils and monocytes to endothelial layers: multiple adhesion and activation steps lead to function of β_2 integrins which provide the necessary adhesive strength, but the specificity comes from the interaction of multiple receptor-counterreceptor pairs and activation steps.[30,31] Strong, antigen-specific adhesion of T lymphocytes to antigen-presenting cells is also mediated by adhesive and activation cascade: weak but specific adhesion via the T cell receptor-CD3 complex triggers activation of β_2 integrins, leading to strong adhesion.[20,114,115] The activation of β_2 integrins in neutrophils, monocytes, and lymphocytes is accomplished by the involvement of protein kinase C in the activation pathways.

The levels of β_1 integrins that are widely expressed on lymphocytes and leukocytes[116,117] increase after antigen stimulation. The β_1 integrins mediate attachment of lymphocytes, well as many other cell types, to extracellular matrix protein and likely play a role in extravasation and migration of activated lymphocytes in tissues during immune responses.[118] Conversion from naive to memory T cells is accompanied by changes in the surface levels of β_1 integrins. The activation mechanisms of β_1 and β_2 integrins in lymphocytes are probably very similar, because activation of T cells by antigen or by phorbol esters leads to activation of β_1 integrins and leads to increased adhesion of activated T cells to collagen, fibronectin, and laminin.[119-121]

Phosphorylation of β_1 integrins may cause apparent inactivation of the receptors. During oncogenic transformation by Rous sarcoma virus pp60^{src}, phosphorylation of a tyrosine residue in the β_1 subunit appears to reduce binding of β_1 integrins to both talin and fibronectin.[122] There is additional evidence supporting the correlation of tyrosine phosphorylation of β_1 integrin with inactivation.[123] The specific serine residue in β_1 integrin becomes phosphorylated in mitotic cells and the β_1 integrin from such cells no longer binds to fibronectin, consistent with the rounding and detachment of cells during mitosis.[20] Thus, inactivation of integrins may be regulated by phosphorylation. The constitutively inactive state of the integrins on platelets and leukocytes is crucially important for control of thrombosis and inflammation. Transitorily, during cell migration or mitosis, attached cells need to detach from extracellular matrix or from other cells. In a variety of cell types, occupation of integrin receptors by their ligands leads to tyrosine phosphorylation and cytoplasmic alkalinization.[20] It has been suggested that the convergence of the signaling pathways triggered by soluble growth factors, extracellular matrix adhesion receptors, and tyrosine kinase oncogene products that could offer a potential explanation for the anchorage dependence of growth of normal cells and its loss in transformed cells.[124] Adhesion of fibroblasts, endothelial cells, and lymphocytes to fibronectin causes elevation of cytoplasmic pH, which correlates with the parallel stimulation of spreading and growth.[125-127] In the case of cytoplasmic alkalinization, as in the case of tyrosine phosphorylation, one can conclude that cell adhesion signals generated by integrins converge and synergize with those triggered by soluble growth factors and oncogenes, because constitutively elevated cytoplasmic pH in transformed cells correlates with anchorage independency of growth.[128] Therefore, in a variety of cell types (platelets, fibroblasts, carcinoma cells, endothelial cells, lymphocytes, monocytes, neutrophils), integrins act as costimulatory signaling receptors, and the integrin-mediated signaling pathways converge and synergize with those due to the soluble ligands.[20]

It has been proposed that clasterization and cross-linking of integrins brings them together, clustering protein tyrosine kinases in a submembranous patch where they become activated and/or react with their substrates.[20] In an adhesion process, integrins provide the strong stable attachment, but only after cell activation by other stimuli such as soluble ligands and mediators as well as interactions with extracellular matrix or other cells. The consequences of activation of specific integrins may include a number of different cellular events: (a) enhancement of cell adhesion; (b) activation of cell proliferation; (c) activation of secretion mechanisms; (d) morphological changes, e.g., cytodifferentiation. Particularly interesting is the concept of integrin-mediated anchorage-dependence of growth.[20] Extracellular matrix proteins acting through integrins stimulate tyrosine phosphorylation inside the cells. These signals may converge or synergize with those from growth factor receptors and, in transformed cells, with oncogene products such as $pp60^{src}$. The requirement of synergy between soluble growth factors and extracellular matrix adhesion in activation of tyrosine phosphorylation and cell multiplication would give anchorage dependence of growth for normal cells. Subversion or replacement of the need for extracellular matrix generated signals by an oncogene tyrosine kinase activation may lead to anchorage independence of growth for transformed cells. The multiple domains of extracellular matrix molecules cluster different cellular receptors to generate combined intracellular signals, analogous to a mechanism generated by clustering receptors in T cells as a result of their interaction with antigen-presenting cells.[20] It has been suggested that the modular structure of extracellular matrix and multiple distinct sites for interaction with cells are designed to cross-link several different surface receptors together in the plane of the membrane, to organize specific cytoskeletal structures, and to trigger and/or generate intracellular signaling.[20] The evidence that fibronectin segments with multiple domains trigger greater tyrosine phosphorylation than do simpler cell-binding domains,[124] and that larger fibronectin domains promote more cytoskeletal organization,[129,130] supports this concept.

Loss of specific integrins can produce reduced adhesion of tumor cells for extracellular matrix. Viral transformation of rodent cells is accompanied by loss of $\alpha_5\beta_1$ integrin and reduced adhesion to fibronectin,[18] and overexpression of transfected $\alpha_5\beta_1$ increases adhesion to fibronectin and reduces tumorigenicity of such cells.[19] However, the acquisition of new adhesive properties by tumor cells has also been observed.[21] Expression of $\alpha_V\beta_3$ integrin correlates well with the invasion of melanomas.[131] The elevation of a collagen/laminin receptor ($\alpha_5\beta_1$ integrin) occurs in tumor cells,[132] and transfection of cDNA for $\alpha2$ into rhabdomyosarcoma cells leads to acquisition of metastatic capacity.[133] The inhibition of cancer cell metastasis[134-137] and invasion[138] by synthetic fibronectin (RGD-containing) and laminin (YIGSR) peptides has been observed.

Recent experimental evidence on adhesion-dependent regulation of programmed cell death (apoptosis) strongly indicates that the expression and function of specific integrins such as $\alpha_V\beta_3$ and $\alpha_5\beta_1$ are correlated with resistance to apoptosis of certain cell types, and disruption of integrin-mediated adhesion induces apoptosis (see next chapter for details). It has been suggested that integrin-mediated control of sensitivity toward apoptosis is one of the key elements of the anchorage-dependency of growth which is disrupted in cancer cells.[168] Therefore, impairment or subversion of physiological integrin functions in cancer cells may be a key factor contributing significantly to the formation of basic fundamental features of the malignant phenotype, such as autonomization of cell multiplication, loss of anchorage dependence of growth, and aberrant adhesion behavior of cancer cells.[164] Integrins and other adhesion receptors play a crucially important role in physiological function through a unique modular structure of extracellular matrix, with precise spatial organization of multiple distinct cell binding sites on extracellular matrix proteins. It seems that this is a requirement for the design of a specific cross-link pattern of several different cellular adhesion receptors in the plane of the membrane, and subsequent generation of specific intracellular signaling pathway. The transglutaminase-mediated intra- and intermolecular cross-linking of extracellular matrix proteins may provide an enzymatic mechanism for supramolecular organization and stabilization of extracellular matrix structure . It has been pointed out that primary amine- and glucose aldehyde-dependent reactions for covalent modification of the primary structure of extracellular matrix proteins may lead to disturbances of the above named mechanism for disorganizing structure and function of extracellular matrix proteins in cancer.[13,40,139,140,164] In this context it is interesting to note that normal morphology of transformed cells can be restored in vitro by addition of fibronectin,[141-143] derived from either normal or malignant cells.[142] On the other hand, gross failure in junctional communication associated with neoplastic transformation would cause uncontrolled cancerous growth.[144]

1.2.3. The Role of Cadherins in Cell Adhesion

Cadherins are a Ca^{2+}-dependent protein family of cell-cell adhesion receptors[51] that present on most cell types. All cells that form solid tissues express on the cell surface some of the cadherins, and each of them mediates homophilic adhesion between cells. About a dozen different cadherins have been discovered[51,145] and cells containing different cadherins segregate from each other to form separate homophilic aggregates.[49-52] Cadherins bind cells to each other by means of their homophilic interactions and each cadherin type binds only to identical cadherins on interacting cells.[52,146,147] At least four subclasses of cadherins have been characterized: E-cadherin (uvomorulin or epithelial cadherin); P-cadherin (placental cadherin); N-cadherin (neuronal cadherin); and L-CAM (liver cell adhesion molecule).[51] Each of the cadherin subclasses has a unique pattern of tissue distribution, and many types of cells may be characterized by the expression of a particular cadherin subclass or set of cadherin subclasses.[49,50] Since many cells coexpress multiple cadherin subclasses in different combinations, this unique coexpression pattern of cadherins may create a wide range of tissue specific adhesive interactions. Both qualitative and quantitative differences in cadherin expression may contribute to cell sorting behavior, because the cell lines expressing different amounts of the same type of cadherin segregate from each other in their mixtures.[52,147] Structurally, cadherins contain four homologous repeats in their extracellular domains. The functional significance of three repeats is still unknown; the most distal repeat contains the binding specific determinants, as has been mapped using blocking monoclonal antibody and site specific mutagenesis.[148]

Cadherins have been considered to be important regulators of morphogenesis and down-regulation of cadherins may be associated with invasion and metastasis.[51] The switching on and off of cadherin expression correlates with a variety of morphogenetic events involving cell aggregation or dissociation during development.[49,50] It has been proposed that a switching of cadherin expression is involved in the segregation of cell layers during morphogenesis and the spatio-temporal regulation of cadherin expression at both quantitative and qualitative levels is essential in regulation of morphogenesis.[49-51] The earliest developmental expression of E-cadherin is in morulae, where it plays a key role in compaction, leading to polarization of the cells of the early blastocyst in the mouse embryo.[49,50] Therefore, tissue segregation during embryogenesis is controlled by regulated changes in expression of cadherins, and loss or diminished expression of tissue specific cadherins will lead to dissociation of cells from coherent tissues to become individual migratory cells.[31,49-51]

The intracellular domain of cadherins is very important functionally and is the most conserved region in this molecular family of adhesion receptors.[51] Partial or complete deletion of this domain leads to the abrogation of cell-cell adhesion function of cadherins, even with the extracellular domain remaining intact.[149-153] Wild-type functional cadherins are concentrated at the contact sites between cells, in association with the cytoskeleton (cortical actin bundles); this association of cadherins with the cytoskeleton is crucial for their cell adhesion function.[154-156] The cadherins in cells are always associated with cytoplasmic proteins termed catenin α, β, and γ[150-152,155] and these catenin-cadherin complexes are associated with cytoskeletal components at sites of cell-cell junctions. The formation of these molecular complexes is prerequisite for the cell adhesion function of cadherins.[50] A number of observations suggest that the extracellular domain of cadherins retains a binding ability and specificity in the absence of the authentic intracellular domain or the cytoplasmic elements.[51] However, the interactions between individual cadherin molecules in the plane of the cell membrane may be too weak to connect the cells strongly. Therefore, the cytoskeleton might support the lateral interactions of cadherins or it might anchor them to the cell-cell contact zone, and thus the cadherins may aggregate laterally on the membrane at the cell-cell contact site to achieve sufficient for strong adhesion cell-cell binding forces.[51] There is an obvious striking parallel between these processes and the suggested role of extracellular matrix proteins in aggregation and cross-linking of integrin adhesion receptors in the plane of membrane during cellular adhesion (see above).

At the ultrastructural level cadherins have been identified at the zonula adherins junctions, possibly in association with other multiple classes of proteins, such as vinculin, α-actinin, radixin, and actin filaments at the cytoplasmic plaques.[157] Cadherins are known to be phosphorylated and the zonula adherins junctions may be the target for action of the *src* proto-oncogene family of protein kinases.[51] As in case of integrin adhesion receptors (see above), tyrosine kinases may regulate the function of catenin-cadherin-cytoskeleton complexes or they may generate and mediate transmission of intracellular signaling.[51] The activity of cadherins influences the formation of tight, gap, and desmosome junctions, suggesting structural and, possibly, functional association among cadherin-bearing junctions and other junctional structures.[158-162] Therefore, on the molecular and time scale model of cellular adhesion, cadherins will participate in the late stage of cell adhesion comprising secondary stable attachment and strong cell-cell binding.

Loss of adhesive function by specific cadherins, or downregulation of their expression on the cell surface will lead to dissociation of cells from coherent tissues; when cadherins are inactive the cells can be freed from their adhesive interactions and migrate out of their parent tissues.[31,51] It has been suggested that suppression of cadherin activity may trigger the release of cancer cells from the primary tumor and thus may initiate invasion and metastasis.[51] A number of studies have shown relationships between loss of cadherins and acquisition of an invasive and metastatic phenotype by cancer cells. An ovarian cancer line with a high spontaneous metastatic potential showed a very low amount of E-cadherin expression as compared with a cell line which had a lower metastatic activity.[163] The inhibition of E-cadherin function by antibodies allows the MDCK kidney epithelial cells to acquire a migratory invasive phenotype.[14] Transformation of cells by tumor viruses leads to loss of E-cadherins and acquisition of invasive properties. A quantitative correlation between loss of E-cadherins and invasiveness has been observed, and transfection of E-cadherin cDNA into these cells blocked their invasive phenotype.[15-17] Suppression of E-cadherin function by expression of antisense RNA rendered cells more invasive.[17] The expression or activity of cadherins is down-regulated in many human carcinomas, and pathological examination of the human cancer tissues suggests that tumors that lose E-cadherin tend to be more invasive.[51] Therefore, the cancer cells' invasion and metastasis are enhanced by the down-regulation of cadherin expression and/or inhibition of cadherin adhesive function.

In summary, the following sequence of events may be proposed as an explanation of the enhancement of cellular adhesion as a consequence of the activation of specific integrins.[164] Clasterization and cross-linking of integrins brings them together, clustering protein tyrosine kinases in a submembrane patch where they become activated and react with their substrates.[20] It has been suggested that the modular structure of extracellular matrix, and the multiple distinct domains of extracellular matrix molecules for interaction with cells, are designed to cross-link several different cell adhesion receptors together in the plane of the membrane in order to organize specific cytoskeletal structures.[20] The organized cytoskeletal components would support the lateral interactions of cadherins, anchoring them to the cell-cell contact zone. The lateral aggregation of cadherins at the cell-cell contact sites will increase the strength of cellular adhesion and enhance cell-cell binding forces. The linkage of cells by means of protein cell adhesion molecules and/or by lectin-carbohydrate binding is a sufficient and necessary event for the extensive expression of junctional communications, particularly for formation of a large number of adherins junctions and gap junctions.[162] Subsequent activation and mediation of specific intracellular signaling pathways, as well as adequate cellular response to the generated signal, are required for development of tight junctions and desmosomes. The role of cadherin-mediated adhesion in regulation of sensitivity toward apoptosis has not been investigated. It will be interesting to determine whether cadherin function may control the cellular response to apoptotic signals in normal cells, as well as whether the loss of cadherin-dependent adhesion would influence the sensitivity to apoptosis of metastatic cancer cells.

1.3. The Role of Carbohydrate Determinants in the Adhesion of Circulating Cells

One of the key features of the cells of the immune system is the rapid transition between adherent and nonadherent states. Three families of adhesion receptors of the immune system have been discovered: the immunoglobulin superfamily; the integrin family; and the selectins. They are involved in adhesive interactions that direct lymphocyte localization and migration, determine lymphocyte homing to different lymphoid organs, and direct neutrophil localization in inflammation.[70,167] Dramatic progress has been made in our understanding of the molecular mechanism of leukocyte-endothelial cell interaction, and the role of specific carbohydrate-lectin recognition and binding in these processes. Here we will briefly summarize the current view concerning the mechanisms which determine the site specificity of leukocyte-endothelium interaction, and the role of the particular glycoepitopes in endothelial cell-leukocyte recognition. A mechanism will be considered that may be responsible for selective expression of certain BGA-related glycodeterminants on specific leukocyte subsets.

1.3.1. The Site Specificity of Leukocyte-Endothelial Cell Recognition, Adhesion and Extravasation

Rapid transition between adherent and nonadherent states (and vice versa) is one of key biological features of the immune cells. To patrol body homeostasis effectively, the cells of the immune system circulate as nonadherent cells in the blood and lymph and migrate as adherent cells through tissues. The immune cells are able to cross endothelial and basement membranes at specific sites to aggregate at a place of infection and/or adhere to cells bearing foreign antigen. In patrolling the body, the lymphocytes leave the blood, migrate through lymphoid organs and other tissues, enter the lymphatics and return to the blood through the thoracic duct. It is important that priming by a specific antigen may alter the surface phenotype, enabling selective recirculation of lymphocytes to the particular type of secondary lymphoid organs (lymph nodes) where the specific antigen was first encountered.[184,185] Lymphocytes in the blood enter lymph nodes by binding to specialized "high" endothelial cells, and "recirculation" or "homing" receptors on lymphocytes have been defined by monoclonal antibodies that block binding of lymphocytes to the high endothelial cells of specific types of lymph nodes.[184-186] Specific binding and immigration of lymphocytes to the high endothelial venules in lymph nodes is a sequential cooperative multistep "cascade" process involving several receptors on the lymphocytes and counterreceptors on the endothelium.[30,69,70]

The recruitment of leukocytes from the blood into the tissues—leukocyte extravasation—is regulated by selective mechanisms of leukocyte-endothelial cell recognition. The physiologic trafficking of lymphocytes and cellular response to tissue damage and inflammation show extraordinary specificity in relation to the tissue site or organ involved.[30] The tissue-specific selective interaction of lymphocyte subsets with high endothelial venules in lymph nodes and specific recruitment of neutrophils in inflammation are the best examples. However, this specific leukocyte-endothelial cell recognition cannot be explained by a lock-and-key model, since individual receptors participate in multiple leukocyte-endothelial cell interactions that are regulated quite independently in vivo.[30,167] A general model of leukocyte-endothelial cell recognition has been proposed in which this process requires at least three sequential events.[30] Interaction is initiated by transient and reversible adhesion through binding of constitutively functional leukocyte adhesion receptors to endothelial cell counterreceptors. Typically, this step requires endothelial cell activation and such adhesion is mediated by lectin-carbohydrate interactions involving leukocyte or endothelial cell selectins and their oligosaccharide ligands.[62-65,68,187] The second event is activation of the leukocyte by a specific chemoattractant or cell contact which activates the secondary adhesion receptors (the integrin or immunoglobulin superfamily member). The function of those adhesion receptors is activation-dependent and their interaction with endothelial counterreceptors results in strong, stable attachments, completing the recognition process.[30]

The initial reversible adhesion, activation, and activation-dependent stable binding may be a common mechanism for many or all specific leukocyte-endothelial cell recognition reactions.[167] The specificity of those interactions is determined by unique combinations of initial adhesion receptor-ligand and activating receptor-ligand pairs, the number and combination of which is the product of several sequential recognition events, each representing a go/no go decision point.[30] A given highly effective receptor-ligand pair participates in several recognition events and this molecular conservatism does not sacrifice specificity of leukocyte-endothelial cell recognition and leukocyte extravasation.

1.3.2. The Specific Carbohydrate-Lectin Interactions as a Key Molecular Event of Leukocyte-Endothelium Recognition and Adhesion

Three families of immune system adhesion receptors have been discovered: the immunoglobulin superfamily; the integrin family; and the selectins. They are involved in adhesive interactions that direct lymphocyte localization and migration, determine lymphocyte homing to different lymphoid organs, and neutrophil localization in inflammation.[30,70,186] Carbohydrate ligands for selectins have recently been discovered: sialosyl-LeX, sialosyl-LeA, and LeX have been identified as recognition structures for ELAM-1 and CD62.[62-65,187] Neutrophils bear LeX both on glycolipids and at the termini of N- and O-linked oligosaccharides.[60,61] Finally, the study of how leukocytes roll on a selectin at physiologic flow rates elegantly shows that rolling on a selectin is a prerequisite for activation-induced adhesion through integrins.[68] Thus, these results clearly demonstrate the role of particular BGA-related glycodeterminants as specific mediators of the early initial step of leukocyte adhesion.

Based on these and other facts, we have suggested[13,38-42,164,188,353] that the processes of thymic education and antigen presentation are accompanied by certain changes in the glycosylation pattern of the cellular membranes. These changes include the presentation of the BGA-related glycoepitopes—in particular, of their cryptic forms. Aberrant glycosylation may occur as a result of binding of foreign peptides to MHC and subsequent changes in glycosylation patterns of peptide-MHC complexes and/or as a result of presentation of peptide-MHC complexes on the cell membrane and exposition of new, previously masked, glycodeterminants. Numerous experimental data are now available that confirm this suggestion for antigen presentation.[353]

Evidence has suggested that T cell recognition of stimulator (antigen-presenting) cells may be influenced by the state of glycosylation of the stimulator cells.[189-194] Recognition of specific carbohydrate determinants, probably associated with HLA antigens on the surface of antigen presenting cells, may play a key regulatory role in function of both MHC class I and MHC class II restricted T cells. Black et al[194] have suggested a regulatory role for asparagine-linked oligosaccharide in the H-2 restricted cytolysis of virus-infected cells. Tunicamycin treatment of P 815 cells before and during infection with vesicular stomatitis virus (VSV) inhibited by about 50% the lysis of infected P 815 cells by VSV-immune H-2 identical killer cells.[194] Furthermore, numerous data indicate the role of oligosaccharides in cellular recognition by thymic lymphocytes:[189] (a) selective removal of cell surface sialic acids from the surface of antigen-presenting cells supports their recognition by MHC class II reactive T cells;[192,193] (b) murine B cell and spleen adherent cell Ia antigens display a different glycosylation pattern;[190] (c) a potential regulatory function has been proposed for carbohydrate side chains on Ia molecules in T cells—discrimination between Ia antigens expressed on allogenic accessory cells and B cells.[191] Sialic acids on cell surface molecules, including MHC, may play a role in antigen presentation, since removal of sialic acids by neuraminidase can restore specific responses to nonresponder antigen-presenting cells, and effective antigen presentation occurs in cells expressing an increased number of histocompatibility antigens with decreased sialylation levels.[195] Cells infected with AIDS virus or cytomegalovirus start to express high levels of LeY and LeX glycoepitopes, respectively.[196,197] This relatively nonspecific mechanism of the presentation of BGA-related glycoepitopes may be involved in the process of homotypic sorting of cells at subsequent stages of formation of immune response. In view of this, natural anticarbohydrate antibodies, identified in blood serum from most healthy individuals,[198] may fulfill a double function: to accomplish the complement-dependent lysis of cells containing high-density BGA-related glycoepitopes (e.g., in thymic education it may lead to negative selection of T cells whose major histocompatibility complexes have a high affinity for their own presented antigens); and to favor the homotypic selection of cells revealing a medium or low density level of the BGA-related glycoepitopes.[13,40,42,164,188,353]

The study of M. Huesmann et al[199] shows that the binding of TCR to thymic MHC molecules rescues $CD4^+8^+$ cells from programmed cell death and induces, first, upregulation of the TCR level and then differentiation into $CD4^+8^-$ or $CD4^-8^+$ cells in the absence of any cell division. The proposed sequence of positive selection of class I MHC-restricted $CD4^-8^+$ T cells in the thymus[199] can be combined with our concept as follows.[13,42,164,188,353] Immature $CD4^+8^+$ thymocytes normally express a large variety of TCRs and are BGA-related glycoepitope positive. Some of the $CD4^+8^+$ cells, whose TCRs can bind to class I MHC ligands on thymic epithelium, are rescued from cell death and selected to mature. Thus, nondividing $CD4^+8^+$ cells with high TCR levels and masked BGA-related glycodeterminants (possibly, by sialylation) are formed. Then they become nondividing $CD4^-8^+$ TCR high mature thymocytes with no expression of BGA-related glycoepitopes on their cell surface. At the stage of negative selection, T cells with high affinity receptors to self peptide-self MHC molecules become positive for BGA-related glycoepitopes, as a result of, for example, transfer of self peptides from cell surface of antigen presenting cells into T cells and subsequent association of self peptides with MHC molecules of corresponding T cells. Thus, the T cells with high, medium, and low level affinity for self peptide-self MHC complexes will have corresponding levels of association of self peptides to their own MHC, and correspondingly high, medium, and low levels of BGA-related glycoepitope expression on their cell surface. Subsequently, T cells with high densities of BGA-related glycoepitopes on cell surfaces, such as immature thymocytes, will be targets for elimination. This "self peptides pick-up mechanism" by T cells provides one of the possible explanations for how antigen-presenting function could participate in negative selection during thymic education and, probably, in positive selection. For example, association of TCRs with ligands could protect the former from proteolytic degradation and lead to TCR up-regulation. Subsequent up-regulation of MHC could be a second checking and selective signal for T cells with high affinities to MHC-self peptide complexes.[164,353]

Desialylation of cell membrane glycomacromolecules could be a major factor in determining the activation of the cell adhesion mechanism. Despite their larger surface areas, T cell blasts and thymocytes have five-fold less sialic acid per cell than resting T cells,[200] and are less negatively charged,[201] which may be important factors determining whether the adhesive mechanisms are active or latent. In lymph nodes, the activated antigen-responsive lymphocytes that aggregate in germinal centers are greatly undersialylated, whereas areas containing B and T cells in rapid transit between blood and lymph are sialylated normally.[202] Polysialylation of NCAM in the nervous system antagonizes its ability to promote adhesion.[32] Desialylation of the cell membrane glycomacromolecules could be a major mechanism in the regulation of negative charge on the cell surface, which is mainly responsible for preventing close cell-cell contract between circulating cells. Thus, a decrease in the sialylation level of the cell membrane glycomacromolecules could be a simple primary mechanism for the activation of cell adhesion. It would be accompanied by appropriate changes in physical properties (neutralization of negative charge) and structural characteristics (expression or exposition of cell adhesion glycodeterminants) of the cell membrane, alteration of the morphology of contact between cells, and activation of latent cell adhesion mechanisms. However, the carbohydrate-mediated cell-cell recognition regulatory mechanism may play an important role beyond the initiation of multistep "cascade" cell adhesion reactions. It has been suggested that the extent of glycosylation on ICAM-1 may regulate the endothelium adhesion to LFA-1 or MAC-1 (the leukocyte integrins),[203] and thus control the stable binding of leukocytes to the endothelium, since this reaction involves the activation-dependent leukocyte integrin MAC-1.[30]

Specific carbohydrate ligands are involved in differential attachment to endothelium and subsequent migration into the tissues of specific lymphocyte subsets. Naive and memory-type T cells show clear differences in their distribution and migration patterns. The memory T cells accumulate in inflammatory lesions and are the only T cells which recirculate through uninflamed tissues including skin.[204] In contrast, the local lymph nodes are designed for the massive migration of lymphocytes of diverse specificities, especially naive T cells. Picker et al,[66] and Shimizu et al,[67] showed that a molecule termed endothelial cell-adhesion molecule-1 (ELAM-1) is one of the receptors on endothelium for a specific subset of circulating T lymphocytes: only memory-type T cells ($CD4^+$ T cells), a functionally potent class of T cells, bind ELAM-1. ELAM-1 was originally implicated in the binding of blood neutrophils and monocyte-like cells to endothelium.[205] Ligands for ELAM-1 expressed on neutrophils are sialylated, fucosylated polylactosamines, such as the sialosyl Lewis X determinants, which bind to the lectin domain of selectins (see above). The carbohydrate ligand for ELAM-1 for T cells has not been identified, but it appears to be highly restricted in its expression to a specific subset of memory T cells, particularly of resting peripheral $CD4^+$ T cells.[67] Since ELAM-1 has been identified as a molecule involved in the activation-independent binding of resting circulating memory T cells to endothelium, this adhesion pathway will allow inflamed endothelium to selectively "capture" that particular subset of lymphocytes. L-selectin is constitutively functional, presents at high levels on circulating, non-activated, resting neutrophils and mediates their attachment to cytokine-stimulated endothelium by presenting neutrophil carbohydrate ligands (e.g., sialyl Lewis X) to the vascular E- and P-selectins.[78] However, in the absence of secondary activation-triggered stable attachment or in the presence of adhesion inhibitor(s), this primary adhesion is reversible and arrested circulating cells will be released and returned to the circulation.[30] Even activation-dependent adhesion via leukocyte integrins, which is stable for minutes under physiologic shear forces, is in principle reversible: neutrophils bound to endothelium by integrin-mediated adhesion (Mac-1) are released spontaneously after 10-15 minutes.[206] Thus, extravasation is not obligatory following even specific stable attachment of leukocytes to endothelium.

Therefore, specific interactions of selectins with corresponding carbohydrate ligands, particularly specific BGA-related glycodeterminants, are essential for the selective attachment of circulating cells such as neutrophils, monocytes, and lymphocytes to endothelium, and their subsequent migration into the tissues—the process which is critically important for the maintenance of homeostasis. Furthermore, interaction of T cells with antigen-presenting cells also involves specific carbohydrate determinants, which are presumably associated with MHC molecules. It will be particularly interesting to determine whether carbohydrate-lectin interactions may influence the sensitivity toward apoptosis of the immunocompetent cells, and whether carbohydrate-mediated adhesion may generate a survival signal for circulating normal cells.

1.4. The Role of Carbohydrate Determinants in Cellular Adhesion During Embryo- and Organogenesis

Glycoantigens undergo dramatic changes during development, cell differentiation and maturation.[207-209] Cancer cells frequently express carbohydrate embryonic antigens since they undergo a retrodifferentiation process during the course of malignant transformation.[210] SSEA-1 (stage-specific embryonic antigen-1), first described by Solter and Knowles in 1978,[211] is one of the embryonic antigens, which is specifically expressed on the murine preimplantation embryos at the morula stage. The SSEA-1 antigen is described chemically as a carbohydrate antigen carrying LeX-hapten and i-antigenic structures.[212,213] By using specific monoclonal antibodies, it was shown that SSEA-1 and its modified form of the antigen, including the sialylated form (sialylated LeX or sialylated LeX-i) and fucosylated forms (LeY and poly LeX), are frequently accumulated in human cancer of various origins.[214-218] Lung adenocarcinoma is one of the good sources of the SSEA-1 and its modified forms of the antigen,[219] and murine monoclonal antibodies directed to lung cancer cells frequently have been found to recognize SSEA-1 and related antigens.[220,221] Moreover, among the modified forms of the SSEA-1, sialylated SSEA-1 has been shown to be present in the sera of patients with adenocarcinoma of the lung, as well as in cancer cells.[217,222,223] Since SSEA-1 was first described as an embryonic antigen and subsequently found in various human cancers, the antigen has sometimes been regarded as an oncofetal or oncodevelopmental antigen.[224-228]

The localization of three carbohydrate antigens, LeX, LeY, and sialylated LeX-i, which are closely related to stage-specific embryonic antigen 1 in the lung of developing human embryos, was investigated using specific monoclonal antibodies.[210] The three antigens all have physiological significance as stage-specific developmental antigens of the human lung; these antigens were specifically present in the bud cells at each important step of the morphogenesis of the human lung, such as cells in the lung buds, bronchial buds, and terminal buds for the formation of the alveolus, and cells differentiating into bronchial gland cells. The three antigens gradually disappear in the later stage of development along with the maturation process of the lung. Stage-specific embryonic antigen 1 and related antigens are known to be associated with various human cancers, including lung cancer. Miyake et al[210] suggest that the expression of these antigens in the lung cancer cells is the result of the retrodifferentiation of the cancer cells to the stages of immature embryonic lung cells.

It was known that SSEA-1 antigens reappear in a certain lineage of cells after implantation during the course of morphogenesis of certain organs such as brain and urogenital organs,[227,229] and stage specific expression of LeX and related antigens has been shown in developing lungs of human embryos.[210] One type of expression of the stage-specific antigen is characterized by strict stage specificity and disappearance of the antigen in mature cells. This expression of SSEA-1-related antigen during the course of organogenesis has been regarded as the second wave of the expression of those antigens,[210] assuming the appearance of these antigens in the preimplantation embryos as classically described[211] as the first wave of the expression. The third type of expression of stage-specific antigen is characterized by the persistence of the antigen after differentiation. This expression of the SSEA-1 related antigens has been regarded as the third and the latest wave of the expression of these antigens in intrauterine life.[210]

Direct evidence for the involvement of LeX antigen in embryonic development has been shown by studying the effects of oligosaccharides on preimplantation-stage mouse embryos. During compaction, blastomeres maximize their intercellular contacts and generate a polar distribution of microvilli. The appearance of SSEA-1 (LeX) on the mouse embryo surface correlates in time with the onset of compaction.[211] It has been suggested and confirmed that the stage-specific appearance of LeX may play a role in stabilizing compaction and in sorting LeX-positive blastomeres to the inner cell mass during blastocyst formation,[44,46,230] and in the attachment of mouse blastocysts on endometrial monolayers.[45] Oligosaccharides purified from human milk were tested in this experimental system, added either as monomers or as multivalent haptens attached to the primary amino groups of lysyllysine.[44,230] Multivalent LeX haptens were found to make a dramatic decompaction of fully compacted 16-cell embryos, whereas multivalent LeA or GlcNac chitobiose haptens did not cause decompaction. It has been suggested that LeX valency has a critical importance, since a multivalent LeX hapten was able to decompact embryos, whereas free oligosaccharide was largely ineffective.[44,230] In many other biological systems carbohydrate recognition requires a high density or clustering of sugar residues. For example, LeX is carried by the early mouse embryo on highly branched lactosaminoglycans (embryoglycans). Lactosaminoglycans are carriers for many stage-specific embryonic antigens, including SSEA-1 (LeX), LeY, I, i, and the ABH histo-blood group system.[44] However, lactosaminoglycans purified from F9 embryonal carcinoma cells caused rapid agglutination of mouse embryos,[44] agglutinate F9 embryonal carcinoma cells and participate in embryonal carcinoma cell adhesion.[231] Thus, the difference in the range of molecular weight of carrier, as well as the density and the valency of oligosaccharide haptens, may cause the opposite biological effect mediated by the same glycodeterminant. It is important that, like many other cases of cell-cell type specific recognition and adhesion, the process of compaction requires the cooperation of multiple cell adhesion molecules and mechanisms. Compaction is known to be controlled by a number of cell surface molecules, including uvomorulin,[232] galactosyl transferase[233] and gap junction proteins.[234] Selective carbohydrate-mediated cell-cell type specific recognition and adhesion play a crucial role in embryogenesis. During development carbohydrate-based adhesion may serve as a mechanism of "sorting-out" of homotypic cell populations from a heterotypic assemblage of cells and, probably, for initiation of secondary, stronger and more stable protein-mediated cellular adhesion. Is carbohydrate-mediated adhesion important for survival of embryonal cells? It will be of interest to investigate whether carbohydrate-lectin interactions may trigger an apoptosis-resistance program in a subset of embryonal cells during development.

1.5. The Cancer-Associated Glycoepitopes and Sugar-Binding Proteins as Key Molecular Determinants in Cancer Adhesion and Metastasis

Invasion and metastasis of cancer cells are complex multistage processes affected by a multiplicity of mechanisms and involving numerous factors.[177,235,236] The ability of tumor cells to detach from the primary lesion, to penetrate through basement membrane surrounding blood vessels, and finally to implant into remote tissues and organs, is mediated in part by hydrolytic enzymes.[179,237] The spectrum of biological activities of the plasminogen activation system makes it an ideal candidate to play a key role in cancer invasion,[237,238] and there is numerous data supporting the hypothesis for a role of the plasminogen activation system in tumor invasion and metastasis.[180,181,239-243] Furthermore, the aberrant adhesion behavior of cancer cells contributes significantly to the metastatic dissemination. The adhesion of transformed cells is characterized by preservation of cellular recognition functions mediated by carbohydrate and peptide cell adhesion epitopes. Inability to display the secondary stable attachment and strong terminal tissue-specific adhesion, with impairment of adequate intracellular signaling and biological cellular responses are also typical for cancer cell adhesive behavior. Carbohydrate-mediated cell-cell and cell-substratum adhesion play a critically important role in cancer metastasis. Even for invasion, the attachment of cancer cells to the basement membrane and extracellular matrix is required. The invasive phenotypic functions include motility, loss of homotypic cell-cell adhesion, establishment of alternative cell-substrate and heterotypic cell-cell adhesion and breakdown of extracellular matrices.[244] Cancer cell adhesion is critically important for arrest of circulating tumor cells at the distant specific site(s) of metastasis.

Blood capillaries may act as a sieve and thus arrest tumor cells with a diameter bigger than that of blood capillaries. However, tumor cells change their shape to pass through narrow capillaries.[245] Lymph node transmigration has been tested by intralymphatic injection of tumor cells. Fisher and Fisher[246] found that different types of carcinoma cells rapidly moved through lymph nodes. Kurokawa[247] has observed that hepatoma cells were retained in the regional lymph nodes and he did not detect tumor cells elsewhere until macroscopic metastases had developed in the regional nodes. Therefore, tumor cells are retained in target organs because they adhere to the other cells and/or to the endothelium of the organ's capillaries. On the basis of circulatory anatomy one could predict where about 60% of the cancer metastases will arise.[29] For this reason the lungs are a common site of metastasis for many human cancers, and metastases from colon carcinoma often develop in the liver.

1.5.1. Cancer Cell Aggregation and Metastasis

Adhesiveness is one of the key cell surface-mediated properties that is altered during malignant transformation,[1] and abnormal adhesiveness of tumor cells is certainly thought to contribute to the ability of cancer cells to metastasize. One of the characteristic features of the growth of malignant tumors cultivated in vitro is their ability to form three-dimensional, spheroid-like structures termed multicell spheroids.[182,183] Similar spheroid-like formations are observed in rodent and human malignant tumors in vivo (according to biochemical, immunocytochemical, morphological, and cytological criteria). Therefore, multicell spheroids in vitro are considered a cytofunctional analog of the avascular stage of in vivo tumor development.[183] Cell association and aggregation occur in the first stage of multicell spheroid formation and can be measured in vitro using the cell-aggregation assay. Many studies indicate that cell surface components play an important role in metastasis,[178,248,249] in particular in formation of tumor cell aggregates and emboli. Circulating metastatic tumor cells tend to adhere to other tumor cells (homotypic adhesion) or to host cells, including platelets and lymphocytes (heterotypic adhesion), to form emboli, the formation of which is correlated with metastasis.[8,250,251] The lodgment, attachment, and growth of blood-borne neoplastic cells depend largely on embolization.[252,253] In a B16 melanoma experimental model of metastasis,[254,255] it has been shown that tumor cell clumps produce more lung metastases after i.v. injection than do single cells[8] and a correlation has been demonstrated between the tendency of cells to undergo homotypic and heterotypic aggregation in vitro and their metastatic potential in vivo.[253,256-259] The concentration and size distribution of tumor cell clumps that enter the circulation play a significant role in the metastatic process.[8-11] Intravenous injected tumor cells in clump form have a greater tendency to form metastases than do the same number of single cancer cells.[8,11,12,251]

On the other hand, among hepatomas, low homotypic adhesion was found to be correlated with high metastatic capacity,[260] and it has been suggested that decrease in homotypic adhesion possibly causes the cancer cell release and dissemination from the primary tumor.[178] It has been shown that reduced cell adhesion to fibronectin and laminin is associated with altered glycosylation of β1-integrins in a weakly metastatic glycosylation mutant of B16 melanoma cells.[261] This mutant showed reduced metastatic capacity and decreased adhesion to endothelial cells and endothelial cell extracellular matrix despite increased homotypic cell adhesiveness.[262,263] It has been suggested that the intrinsically low adhesiveness of cancer cells contributes to separation of individual cells or small groups of cells from the primary tumors.[2] Coman and others[3-5] have observed that tumor cells were more easily separated from solid tumors than were normal cells from corresponding tissues. It has been shown that tumor cells were less adherent than normal cells to artificial substrates.[6,7] The mutant cell lines of B16 melanoma cells of mouse with reduced metastasizing ability and an increased cell-to-cell adhesive property contain elevated concentrations of α(1-3) fucosyltransferase and an increased surface expression of the SSEA-1 determinant.[299]

A number of studies have demonstrated differences in cell surface glycoconjugates between tumor cells exhibiting high and low metastatic capacity. It is quite well established that cell surface constituents are involved in cell-cell and cell-substratum recognition and adhesion[264] and that such interactions are relevant for tumor cell embolization, specific arrest, and implantation in secondary sites.[178,248,249] The emboli may lodge nonspecifically in narrow capillaries, or they may adhere preferentially in specific organs, presumably as a result of cognitive interactions between the tumor cells and capillary endothelial cells or exposed basement membrane components.[178,248,249] Tumor cell surface lectins play major roles in cellular interactions in vivo that are important for the formation of emboli and for the arrest of circulating tumor cells.[265-269] Antilectin antibodies localize lectins in the cytoplasmic compartment of various normal cells as well as in the extracellular compartment, where they seem to be associated with the extracellular matrix after being secreted by the cells.[270-274] Various functions have been proposed for the lectins, including mediation of intercellular recognition and adhesion,[274-280] control of cellular proliferation,[270,281,282] organization of the extracellular matrix,[270,272,274] and elastic fiber formation.[283] Tumor cells and tissues contain lectins that are similar to those isolated from normal cells in their sugar-binding specificities, molecular weights, and sequences.[269,284-292] It was found by using monoclonal and polyclonal antilectin antibodies that the lectins were expressed on the cell surface and that their levels correlated with malignant transformation and metastatic propensity.[288,289,291] A key role has been suggested for the tumor-surface lectins in anchorage-independent growth in vitro and in tumor embolization (homotypic and heterotypic aggregation) during the metastatic process in vivo.[265-269] A number of rodent and human malignant tumors express lectins with biospecificity to galactose, lactose, and N-acetyllactosamine.[279,286,293,294] It has been considered, and confirmed for quite some time, that adhesion of platelets to tumor cells is a prerequisite for cancer metastasis.[137,257,295-298]

The homotypic aggregation of B16 melanoma cells is dependent on the presence of fetal bovine serum.[253,259] One possible explanation for this serum requirement could be that a serum glycomacromolecule(s) mediates intercellular adhesion similarly to the action of cell-cell adhesion molecules. Our studies have shown that the serum glycoamines and other macromolecules may play important roles in changing the adhesion and aggregation properties of cancer cells in vitro and metastasis in vivo and thus may be involved in modification of cell recognition, association and aggregation.[13,38-42,164] Since carbohydrates and carbohydrate-recognition structures are involved in cell aggregation, we have initiated the investigation of structural-functional interrelations of glycoamines, their synthetic structural analogs, and high molecular weight carbohydrate-associated serum proteins as modifiers (inhibitor or stimulator) of tumor cell aggregation, in vitro and metastasis in vivo. The results of ongoing research programs[13,40,42,170,171] have provided increasing evidence to support the hypothesis that: (a) different molecular isoforms of glycomacromolecules in serum contain cell-originated glycoepitopes which are involved in cell recognition and cell association processes (the family of cell recognition glycoepitopes containing extracellular biopolymers (biomacromolecules)—CRG-CEB); (b) these different molecular isoforms of the CRG-CEB and serum carbohydrate-binding proteins (e.g., anti-carbohydrate antibodies, lectins, etc.) play opposite roles, either as inhibitors and/or stimulators of cell association and aggregation, depending on the nature and size of the carrier molecules, valency of glycoepitopes (for CRG-CEB) and binding properties (for carbohydrate-binding proteins); (c) human cancer is accompanied by changes in quantitative and qualitative characteristics of the CRG-CEB and carbohydrate-binding proteins in blood serum; and (d) benign and malignant human tumors have different in vitro cell aggregation properties.

Thus, the homotypic and heterotypic aggregation properties of tumor cells represent important biological features of malignancy because these properties of transformed cells could determine the metastatic ability of neoplastic tumors.[8,178,248-253,256,257,259,265-269] The key structural determinants of the tumor cells that participate in cell recognition, association and aggregation have been determined as carbohydrates and carbohydrate-binding proteins.[178,248,249,264-270,279] The acquisition of certain adhesive properties by malignant cells is very important for invasion, motility and metastasis.[12,13,21] On the other hand, the profound defects in protein adhesive systems primarily mediated by cadherins[14-17] and integrins[18-21] is characteristic of malignant transformation and would contribute significantly to the abnormal locomotion, motility, invasion and metastatic behavior of cancer cells. Therefore, a number of different subpopulations of cancer cells with different adhesive properties would play a significant role at certain stages of the multistage process of metastatic spreading, and metastatic dissemination should require both the cooperative interactions among the cancer cell subpopulations and involvement of normal cells.

The biology of cancer cells cannot be described and understood as a simple complex or mixture of the characteristics of their component cell populations, both neoplastic and normal.[300] The malignant behavior of tumor tissue arises as a result of multiple cellular interactions among diverse cell populations comprising cancers as "cell societies".[301] The mixture of different tumor cell subpopulations is subject to interactions among their component and these interactions significantly modify growth, drug resistance and even the metastatic ability of individual cancer cell subpopulations.[302-304] Apparently nonmetastatic tumor cells and nonmalignant cells, such as embryonic cells, may participate in metastasis if mixed with metastatic cells, by increasing the number of metastatic nodules that result from i.v. injection of metastatic cancer cells.[8,304,305] Nonmetastatic cancer cells have the ability to metastasize in the presence of metastatic cells in vivo;[304] this has been confirmed by the recovery of clonogenic cells of a nonmetastatic variant from metastatic lung nodules.[304,305] Therefore, cellular interactions among normal and cancer cell subpopulations may allow metastases to be established, even by tumor cells that are independently deficient at individual steps of the metastatic cascade.[300]

Generally, tumor cells metastasize to the organ containing the first capillary bed encountered after leaving the primary tumor; however, other tumors exhibit preferential metastasis to specific organs: ocular melanoma metastasizes to the liver, and clear cell carcinoma of the kidney spreads to the thyroid.[248,306] It is clear that tumor cells do not play an exclusive role in metastatic spreading, but some other cellular counterparts (e.g., leukocytes, platelets) are also directly involved in metastasis. There is a problem in obtaining metastasis by injecting human tumor cell lines in nude mice: most do not metastasize in nude mice, even if originally derived from a metastatic lesion.[307] It is possible that some of the important host cellular metastatic partners are missing in this experimental system. Recent studies indicate that normal tissue influences metastasis such that many human and animal tumors transplanted into the nude mice metastasize only if placed in the orthotopic site.[308] It is generally accepted that metastases do not occur as a result of random survival of cancer cells released from the primary tumor but from selective growth of specialized subpopulations of highly metastatic cells with certain specific properties.[326] However, these "primary" or "key" metastatic cells would need support from the "secondary" or "helper" cells, both nonmetastatic cancer and normal, in order to complete each step of the metastatic process.

1.5.2. The Tumor-Associated Glycoepitopes and Lectins as Key Molecular Determinants of Cancer Cell Aggregation, Adhesion, and Metastasis

Aberrant glycosylation of cell membrane macromolecules is one of the universal attributes of malignant tumors. A rather limited number of molecular probes based on monoclonal anticarbohydrate antibodies now enables the detection of over 90% of the most widespread forms of human cancer.[77,309] The most characteristic manifestation of aberrant glycosylation of cancer cells is neosynthesis (or ectopic synthesis), the synthesis of incompatible antigens, and incomplete synthesis (with or without the accumulation of precursors) of the BGA-related glycoepitopes.[77,207,309] As we have mentioned earlier, BGA-related glycoepitopes are directly involved in the homotypic (tumor cells, embryonal cells) and heterotypic (tumor cells-normal cells) formation of cellular aggregates (e.g., Lewis X antigens, polylactosamine sequences; and T-and Tn-antigens), which has been demonstrated in different experimental systems.[36,43-46,230] BGA-related alterations in tissue glycosylation patterns are detected in benign (premalignant) tumors with high risk of malignant transformation, in primary malignant tumors, and in metastases,[224,310-314] i.e., they were demonstrated as typical alterations in different stages of tumor progression.

Recently, experimental evidence has generalized that some of the BGA-related glycodeterminants, which had been identified earlier as tumor-associated carbohydrate antigens (TACA), function as key adhesion molecules.[13,42,53,54] Furthermore, the expression of some TACA (sialosyl-LeX; sialosyl-Tn; H/LeY/LeB) distinguishes metastatic deposits from the primary tumor and indicate markedly poorer prognosis in cancer patients with colonic, ovarian, and lung cancer[315,316] (reviewed in ref. 53, 166). It has been suggested that tumor cell populations expressing particular carbohydrate antigens (sialosyl-Tn, sialosyl-LeX H or LeY) may preferentially initiate invasion and metastatic spreading by an unknown mechanism.[53] Earlier the same conclusion had been considered relative to the other carcinoma-associated carbohydrate antigens: T- and Tn-antigens[55,56] and Galα(1-3)Gal epitope.[317] The common cell-biological property of those glycoantigens is their function as key adhesion molecules that initiate a multistep "cascade" cell adhesion mechanism; they are responsible for homotypic and heterotypic cancer cell aggregation. Most recent evidence shows that expression of these glycoantigens may contribute to the increased resistance toward apoptosis of certain subpopulations of cancer cells, particularly metastatic cancer cells.

Therefore, like embryonal and circulating cells, cancer cells do employ specific carbohydrate-mediated recognition mechanisms in several cell-cell type specific homo- and heterotypic adhesive interactions that may ultimately lead to metastatic dissemination. Because, for cells of coherent tissues of the adult organism, these carbohydrate-based cellular adhesion reactions have not been observed so far, we could describe this phenomenon as an acquisition by cancer cells of certain cell adhesion mechanisms that are typical for embryonal and circulating cells. The antigen-presenting function has been proposed as a biological driving force that allows certain subpopulations of cancer cells to acquire these specific adhesive properties as an inevitable consequence of immunoselection and tumor progression.[13,40,42] Recent experimental evidence has confirmed the importance of carbohydrate-mediated cancer cell adhesion in metastatic spreading in vivo. In an experimental model of metastasis it has been observed that specific carbohydrate units of glycosphingolipids, which are involved in cancer cell adhesion, as well as glycosphingolipid-containing lyposomes, significantly suppressed B16 melanoma lung metastasis.[318] Our recent experimental observations also strongly support this concept. Synthetic glycoamines inhibit the incidence of human breast cancer metastasis in nude mice and suppress colony-forming ability in agarose of highly metastatic human melanoma and breast carcinoma cells.[170,171] Interference with carbohydrate-mediated aggregation and adhesion of metastatic cancer cells causes induction of apoptosis in target cells and increases the sensitivity of cancer cells to chemotherapeutic agents.[170,171,174-176] Thus, carbohydrate-lectin interactions may trigger an important intracellular protective mechanism that metastatic cancer cells utilize to survive in circulation and to develop a higher resistance toward apoptosis.

1.6. Conclusion: A Prospect of Anti-Cell Adhesion Therapy For Cancer Metastasis

The concept of "anti-adhesion" or "anti-aggregation" therapy of cancer, particularly cancer metastasis, has recently been proposed[13,40,42,53,54,164,331] with emphasis on inhibition of carbohydrate-mediated tumor cell aggregation and adhesion. As we have earlier summarized,[13,40,42,164,331] histogenetically different normal (embryonal, lymphoid), and a wide range of cancer, cells and tissues express either transitorily (normal) or permanently (cancer) the definite BGA-related glycodeterminants that are involved in the initial early stages of a multistep "cascade" cell adhesion mechanism. Like embryogenesis, tissue repair, regeneration and remodeling could be accompanied by expression of BGA-related glycoepitopes at definite stages of cell differentiation when "sorting-out" behavior of one homotypic cell population from a heterotypic assemblage of cells occurs. Different molecular isoforms of glycomacromolecules in serum contain, in biologically active form, the cell-originated glycodeterminants which are involved in cell recognition, association and aggregation. These glycomacromolecules and serum carbohydrate-binding proteins (e.g., lectins, anticarbohydrate antibodies, etc.) play opposite roles as inhibitors (low molecular weight glycoamines) and/or stimulators (high molecular weight multiglycoepitope-containing serum macromolecules and naturally occurring anticarbohydrate antibodies) of cell aggregation.

Initial observations have shown the effectiveness of these types of synthetic and natural molecules with biospecificity for glycodeterminants of cell adhesion as inhibitors of in vitro cancer cell aggregation and potential antimetastatic agents in vivo.[13,38-46,53-56,139,230,318-323,332-340] It has been shown, for example, that specific carbohydrate units of glycosphingolipids, which are involved in cancer cell adhesion, as well as glycosphingolipid-containing liposomes, significantly suppressed B16 melanoma lung metastasis.[318] Effective inhibition of an experimental metastasis has been shown for peptide cell adhesion determinants[134-137] as well as for monoclonal antibodies against lectins that participate in tumor cell aggregation.

Dramatic progress has been achieved in application of carbohydrate-based cancer cell adhesion inhibitors for treatment of prostate cancer metastasis in rat[169] and human breast cancer metastasis in nude mice.[170] Thus, the concept of anti-cell adhesion therapy of cancer metastasis contains two important practical new elements: (a) the tumor cell aggregation glycodeterminants and corresponding lectins as target structures for antimetastatic drug development; and (b) a new family of low molecular weight synthetic compounds with biospecificity for glycodeterminants of tumor cell aggregation as a tool for antimetastatic cancer therapy.

The one obvious limitation of this kind of cancer therapy has to be considered. Cancer is characterized by profound disturbances of cell adhesion, particularly of those stages that involve the protein adhesion systems (extracellular matrix proteins, integrins, immunoglobulin superfamily, cadherins and junctional communications, etc.). Aberrations in cancer cell adhesion are reversible, since normal morphology and/or biology of transformed cells in vitro can be restored by addition of fibronectin,[141-143] derived either from normal or transformed cells[142] or by transfection of cadherin cDNA.[15-17] Overexpression of transfected $\alpha_5\beta_1$ integrin increases adhesion to fibronectin and reduces tumorigenicity of transformed cells.[19] It seems that this fundamental basic property of cancer cells is critically important for invasion, the dissemination of cancer cells from the primary tumor and their intravasation. Associated with neoplastic transformation, gross failure in junctional communication among the cancer cells is one of the key factors that probably cause uncontrolled cancerous growth.[144] It has been suggested that decrease in homotypic adhesion between the transformed cells may support cancer cell release from the primary tumor and metastatic dissemination.[178] Therefore, carbohydrate-based anti-adhesion therapy of cancer metastasis ideally should be combined with adhesion-modification therapy designed to restore altered protein-mediated adhesion of cancer cells (see Preface).

Human cancer is accompanied by changes in quantitative and qualitative characteristics of the serum macromolecules, generally in the direction of excessive accumulation of the activities that support tumor cell aggregation in blood vessels and metastasis development.[13,40,42,139] These conditions specifically are changed by surgical elimination of the primary tumor because, as we have shown,[139] in the early period after surgery (2-3 days), the level of glycoamines (the major cell aggregation and metastasis inhibitory component in serum) decreases dramatically in serum of oncological patients. The changes in the levels of stimulatory macromolecules are supposedly less radical, since some of them (anticarbohydrate antibodies) are present permanently in the serum of all healthy individuals,[55,198] and it is well known that high molecular weight proteins in serum have a relatively long half-life. Application of cancer cell adhesion inhibitor(s) for treatment of cancer metastasis should be designed as an additional therapy following surgical elimination of the primary tumor, radiation and/or chemotherapy with emphasis on prevention of homotypic and heterotypic cancer cell aggregation in the blood vessels, extravasation of cancer cells, formation of metastatic deposit(s) and secondary tumor foci in target organs.

L-selectin is constitutively functional, present at high levels on circulating, nonactivated, resting neutrophils, and mediates their attachment to cytokine-stimulated endothelium by presenting neutrophil carbohydrate ligands (e.g., sialyl Lewis X) to the vascular E- and P-selectins.[78] Similarly, the corresponding BGA-related glycodeterminants that are present constitutively on the surface of cancer cells may mediate their adhesion to the endothelium and/or attachment at specific site(s) with leukocytes or platelets. However, in the absence of secondary activation-triggered stable attachment, or in the presence of adhesion inhibitor(s), this primary adhesion is reversible and arrested circulating cells will be released and returned to circulation.[30] Even activation-dependent adhesion via leukocyte integrins, which is stable for minutes under physiologic shear forces, is in principle reversible: neutrophils bound to endothelium by integrin-mediated adhesion (Mac-1) are released spontaneously after 10-15 minutes.[206] Thus, extravasation is not obligatory following even specific stable attachment of leukocytes to endothelium, and this may allow the time necessary for anti-cell aggregation action of a prospective metastasis inhibitor.

The number of tumor cells in the lungs declined very rapidly after i.v. injection: after 24 h 90-99% hd disappeared[22-27] and after 3 days generally less than 1% remained.[23,26,27] This decline is probably not due to a major redistribution of tumor cells; rather the cancer cells appear to degenerate rapidly.[23,28] Death of tumor is not due to contact with immune host cells. Possibly tumor cells are damaged by forces exerted by the blood stream,[28] or other factors may cause a rapid degeneration of cancer cells in blood. Thus, the exceedingly low survival of tumor cells in the circulation is one of the most important aspects of the metastatic process.[28,178] For example, a small difference in cell death (less than 1%) can lead to a large difference in metastasis formation (10 times or more).[178] Therefore, inhibition of extravasation of cancer cells, blocking their homotypic and heterotypic adhesion, would prevent escape of metastatic cells from blood into the tissues and may cause a dramatic reduction or even complete prevention of metastases development.

It has been known for many years that even if the bloodstream teams with cancer cells, there may be no clinical or pathohistological evidence of metastatic dissemination into the target organs.[341-343] The levels of metastatic efficiency with respect to lung colony formation, which refer only to the postintravasation phases of metastasis, were only 0.2% and 0.003% in high and low metastatic variants of B16 melanoma cells, respectively, injected at 10^5 cells into the tail veins.[327,344] There is a rapid phase of intramicrovascular (postintravasation) cancer cell death, completed in <5 min, which accounts for 85% of arrested cancer cells; this is followed by a slow phase of cell death, which accounts for the vast majority of the remainder.[328,345,346] Mechanical trauma is an important factor contributing to the rapid death of the majority of cancer cells arrested in the microvasculature of a number of organs during metastasis.[327,345-347] Weiss[348,349] and Weiss and Schmid-Schonbein[350] have suggested that mechanical trauma is a consequence of shape transitions which must occur when approximately spherical cancer cells enter and move along capillaries having smaller diameters than cancer cells themselves.

One of the intriguing conclusions from the presented concept is that prospective leukocyte-endothelial cell adhesion inhibitor(s) which could be designed as an anti-inflammatory agent may well be an effective antimetastatic and anti-AIDS drug, and vice versa. The simple first step for experimental verification of the above concept is also in the hands of vascular cell biologists: the monoclonal antibody that inhibits endothelial cell-leukocyte recognition and adhesion should also block in vivo cancer metastasis, and cancer cells may preferentially accumulate in vivo at the site of inflammation. The enhancement of distant metastases at the site of inflammation or injury[325,329,330,351,352] or at the site of surgical incisions, even those far from the place of tumor resection[324] is in agreement with the suggested conductor-like role of specific leukocyte subsets in metastasis, since leukocytes accumulate selectively at the site(s) of injury or inflammation.

Among four prospective approaches for anti-cell adhesion (ACA) therapy of cancer metastasis (see Introduction), it seems more promising to use, not antibodies and/or lectins as tools of anti-cell adhesion therapy of cancer metastasis, but glycoepitopes themselves linked to the appropriate low molecular weight carrier for clusterization and to increase bioaffinity[13,40,42] or down-regulator of lectin function. The effectiveness of these types of synthetic and natural molecules as inhibitors of cancer cell aggregation in vitro and potential therapeutic agents in vivo has been recently suggested and reported.[13,38-46,53-56,139,164,166,169-173,230,318-323,332-340]

References

1. Poste G, Weiss L. Some considerations on cell surface alteration in malignancy. In: Weiss L, ed. Fundamental Aspects of Metastasis. Amsterdam: North-Holland, 1975:108525-47.
2. Weiss L. Cell detachment and metastasis. In: Stansly PG, Sato H,

eds. Cancer Metastasis, Gann monograph on cancer research, No. 20. Baltimore: University Park Press, 1977:25-36.
3. Coman DR. Mechanisms responsible for the origin and distribution of blood-borne tumor metastases: A review. Cancer Res 1953; 13:397-404.
4. McCutcheon M, Coman DR, Moore MB. Studies on invasion in cancer: Adhesiveness of malignant cells in various human adenocarcinomas. Cancer 1948; 1:460-467.
5. Tjernberg B, Zajicek J. Cannulation of lymphatics leaving cancerous nodes in studies on tumor spread. Acta Cytol 1965; 9:197-202.
6. Sanford K, Barker BE, Woods MW, Parshad R, Law LW. Search for "indicators" of neoplastic conversion in vitro. J Natl Cancer Inst 1967; 39:705-733.
7. Pastan I, Johnson GS. Cyclic AMP and the transformation of fibroblasts. Avd Cancer Res 1974; 19:303-330.
8. Fidler IJ. The relationship of embolic homogeneity, number, size and viability to the incidence of experimental metastasis. Eur J Cancer 1973; 9:223-227.
9. Knisely WH, Mahaley MS. Relationship between size and distribution of "spontaneous" metastases and three sizes of intravenously injected VX2 carcinoma. Cancer Res 1958; 18:900-905.
10. Nadel EM. Symposium of tumor cells in the circulating blood. Acta Cytol 1965; 9:3.
11. Watanabe S. Metastasizability of tumor cells. Cancer 1954; 7:215.
12. Liotta LA, Kleinerman J, Saidel GM. Quantitative relationship of intravascular tumor cells, tumor vessels, and pulmonary metastases following tumor implantation. Cancer Res 1974; 34:997-1004.
13. Glinsky GV. The Blood Group Antigen-Related Glycoepitopes: Biological Functions and Implications in Immunogenesis, Cancer and AIDS Pathogenesis. Austin, TX: Medical Intelligence Unit, RG Landes Company, 1992:114.
14. Behrens J, Mareel MM, VanRoy FM, Birchmeier W. Dissecting tumor cell invasion: epithelial cells acquire invasive properties after the loss of uvomorulin-mediated cell-cell adhesion. J Cell Biol 1989; 108:2435-2447.
15. Frixen UH, Behrens J, Sachs M, Eberle G, Voss B, Warda A, Lochner D, Birchmeier W. E-cadherin-mediated cell-cell adhesion prevents invasiveness of human carcinoma cells. J Cell Biol 1991; 113:173-185.
16. Chen W, Obrink B. Cell-cell contacts mediated by E-cadherin (uvomorulin) restrict invasive behavior of L-cells. J Cell Biol 1991; 114:319-327.
17. Vleminckx K, Vakaet L, Jr., Mareel M, Fiers W, VanRoy F. Genetic manipulation of E-cadherin expression by epithelial tumor cells reveals an invasion suppressor role. Cell 1991; 66:107-119.
18. Plantefaber LC, Hynes RO. Changes in integrin receptors on oncogenically transformed cells. Cell 1989; 56:281-290.
19. Giancotti FG, Ruoslahti E. Elevated levels of the $a_5\beta_1$ fibronectin re-

ceptor suppress the transformed phenotype of Chinese hamster ovary cells. Cell 1990; 60:849-859.
20. Hynes RO. Integrins: versatility, modulation and signaling in cell adhesion. Cell 1992; 69:11-25.
21. Hynes RO, Plantefaber LC. Integrin receptors for extracellular matrix and their involvement in oncogenic transformation. In: Brugge J et al, eds. Origins of Human Cancer: A Comprehensive Review. Cold Spring Harbor Laboratory, Cold Spring Harbor 1991:293-307.
22. Hewitt HB, Blake A. Quantitative studies of translymphnodal passage of tumour cells naturally disseminated from a non immunogenic murine squamous carcinoma. Br J Cancer 1975; 31:25-35.
23. Fidler IJ. Metastasis: quantitative analysis of distribution and fate of tumor emboli labeled with ^{125}I-5 iodo-2'-deoxyuridine. J Natl Cancer Inst 1970; 45:773-782.
24. Proctor JW, Auclair BG, Rudenstam CM. The distribution and fate of blood-borne ^{125}IUdR-labelled tumour cells in immune syngeneic rats. Int J Cancer 1976; 18:255-262.
25. Proctor JW. Rat sarcoma model supports both soil seed and mechanical theories of metastatic spread. Br J Cancer 1976; 34:651-654.
26. Weston BJ, Carter RL, Eastry GC, Connell DI, Davies AJC. The growth and metastasis of an allografted lymphoma in normal, deprived and reconstituted mice. Int J Cancer 1974; 14:176-185.
27. Kodama M, Kodama T. Enhancing effect of hydrocortisone on hematogenous metastasis of Ehrlich ascites tumor in mice. Cancer Res 1975; 35:1015-1021.
28. Roos E, Dingemans KP. Mechanisms of metastasis. Biochim Biophys Acta 1979; 560:135-166.
29. Liotta LA. Cancer Cell Invasion and Metastasis. Scientific American 1992; Feb:54-63.
30. Butcher EC. Leukocyte-endothelial cell recognition: three (or more) steps to specificity and diversity. Cell 1991; 67:1033-1036.
31. Hynes RO, Lander AD. Contact and adhesive specificities in the association, migrations, and targetting of cells and axons. Cell 1992; 68:303-322.
32. Rutishauser U, Acheson A, Hall AK, Mann DM, Sunshine J. The neural cell adhesion molecule (NCAM) as a regulator of cell-cell interactions. Science 1989; 240:53-57.
33. Bixby JL, Pratt RS, Lillien J, Reichardt LF. Neurite outgrowth on muscle cell surfaces involves extracellular matrix receptors as well as Ca2+-dependent and -independent cell adhesion molecules. Proc Natl Acad Sci USA 1987; 84:2555-2559.
34. Lindner J, Zinser G, Werz W, Goridis C, Bizzini B, Schachner M. Experimental modification of postnatal cerebellar granule cell migration in vitro. Brain Res 1986; 377:298-304.
35. Chang S, Rathjen FG, Raper JA. Extension of neurites on axons is impaired by antibodies against specific neural cell surface glycoproteins. J Cell Biol 1987; 104:355-362.
36. Eggens I, Fenderson B, Toyokuni T, Dean B, Stroud M, Hakomori

SI. Specific interaction between Le^x and Le^x determinants: A possible basis for cell recognition in preimplantation embryos and in embryonal carcinoma cells. J Biol Chem 1989; 264:9476-9484.
37. Kojima N, Hakomori SI. Specific interaction between gangliotriaosylceramide (Gg3) and sialosyllactosylceramide (GM3) as a basis for specific cellular recognition between lymphoma and melanoma cells. J Biol Chem 1989; 264:20159-20162.
38. Glinsky GV. Immunoselective hypothesis of tumor progression. Role aberrant glycosylation, anti-carbohydrate antibodies, extracellular glycomacromolecules and glycoamines. J Tumor Marker Oncology 1990; 5:206.
39. Glinsky GV. Glycoamines, aberrant glycosylation, and cancer: a new approach to the understanding of molecular mechanism of malignancy. In: Molecular Oncology. Oncodevelopment proteins and clinical applications. XVIIIth meeting of the International Society for Oncodevelopmental Biology and Medicine. Abstract Book, Moscow, USSR, September 23-27, 1990:7.
40. Glinsky GV. Glycoamines: Structural-functional characterization of a new class of human tumor markers., In: Sell S, ed. Serological Cancer Markers. Totowa, NJ: Humana Press, 1992:233-260.
41. Glinsky GV, Semyonova-Kobzar RA, Berezhnaya NM. Modification of cellular adhesion, metastasizing and immune response by glycoamines: implication in the pathogenetical role and potential therapeutic application in tumoral disease. J Tumor Marker Oncology 1990; 5:231.
42. Glinsky GV. The blood group antigens (BGA)-related glycoepitope. A key structural determinants in immunogenesis and cancer pathogenesis. Crit Rev Oncol/Hemat 1992; 12:151-166.
43. Fenderson BA, rews PW, Nudelman E, Clausen H, Hakomori SI. Glycolipid core structure switching from globo to lacto- and ganglioseries during retinoic acid-induced differentiation of TERA-2-derived human embryonal carcinoma cells. Dev Biol 1987; 122:21-34.
44. Fenderson BA, Eddy EM, Hakomori SI. Glycoconjugate expression during embryogenesis and its biological significance. BioEssays 1990; 12:173-79.
45. Lindenberg S, Sundberg K, Kimber SJ, Lundblad A. The milk oligosaccharide, lacto-N-fucopentaose l, inhibits attachment of mouse blastocysts on endometrial monolayers. J Reprod Fert 1988; 83:149-158.
46. Bird JM, Kimber SJ. Oligosaccharides containing fucose linked α (1-3) and α (1-4) to N-acetylglucosamine cause decompaction of mouse morulae. Dev Biol 1984; 104:449-460.
47. Kojima N, Hakomori S. Cell adhesion, spreading, and motility of GM3-expressing cells based on glycolipid-glycolipid interaction. J Biol Chem 1991; 266:17552-17558.
48. Kojima N, Hakomori S. Synergistic effect of two cell recognition

systems: glycosplingolipid-glycosphingolipid interaction and integrin receptor interaction with pericellular matrix protein. Glycobiology 1991; 1:623-630.
49. Takeichi M. The cadherins: cell-cell adhesion molecules controlling animal morphogenesis. Development 1988; 102:639-655.
50. Takeichi M. Cadherins: a molecular family important in selective cell-cell adhesion. Annu Rev Biochem 1990; 59:237-252.
51. Takeichi M. Cadherin cell adhesion receptors as a morphogenetic regulator. Science 1991; 251:1451-1455.
52. Nose A, Nagafuchi A, Takeichi M. Expressed recombinant cadherins mediate cell sorting in model systems. Cell 1988; 54:993-1001.
53. Hakomori, S.-I. Possible functions of tumor-associated carbohydrate antigens. Curr Opin Immunol 1991; 3:646-653.
54. Hakomori, S.-I. Possible new directions in cancer therapy based on aberrant expression of glycosphingolipids: Anti-adhesion and ortho-signalling therapy. Cancer Cells 1991; 3:461-470.
55. Springer GF. T and Tn, general carcinoma autoantigens. Science 1984; 224:1198-1206.
56. Springer GF, Cheinsong-Popov R, Schirrmacher V, Desoi PR, Tegtmeyer H. Proposed molecular basis of murine tumor cell-hepatocyte interaction. J Biol Chem 1983; 258:5702-5706.
57. Clausen H, Hakomori SI. ABH and related histo-blood group antigens; Immunochemical differences in carrier isotypes and their distribution. Vox Sang 1989; 56:1-20.
58. Lloyd KO. Blood group antigen expression in epithelial tumors: influence of secretor status., In: Reading ChL, Hakomori S-I, Markus DM, eds. Altered Glycosylation in Tumor Cells. New York: Alan R. Liss, 1988:235-243.
59. Stein R, Chen S, Grossman W, Goldenberg DM. Human lung carcinoma monoclonal antibody specific for the Thomsen-Friedenreich antigen. Cancer Res 1989; 49:32-37.
60. Fukuda M, Spooncer ES, Oates JE, Dell A, Klock JC. Structure of sialylated fucosyl lactosaminoglycan isolated from human granulocytes. J Biol Chem 1984; 259:10925-10935.
61. Symington FW, Hedges DL, Hakomori, S.-I. Glycolipid antigens of human polymorphonuclear neutrophils and the inducible HL-60 myeloid leukemia line. J Immunol 1985; 134:2498-2506.
62. Larsen E, Palabrica T, Sajer S, Gilbert GE, Wagner DD, Furie BC, Furie B. PADGEM-dependent adhesion of platelets to monocytes and neutrophils is mediated by a lineage-specific carbohydrate, LNF III (CD15). Cell 1990; 63:467-474.
63. Phillips ML, Nudelman E, Gaeta FCA, Perez M, Singhal AK, Hakomori S, Paulson JC. ELAM-1 mediates cell adhesion by recognition of a carbohydrate ligand, sialyl-Lex. Science 1990; 250:1130-1132.
64. Polley MJ, Phillips ML, Wagner EA, Nudelman E, Singhal AK,

Hakomori S, Paulson JC. CD 62 and endothelial cell-leukocyte adhesion molecule 1 (ELAM-1) recognize the same carbohydrate ligand, sialyl-Lewis X. Proc Natl Acad Sci USA 1991; 88:6224-6228.
65. Berg EL, Robinson MK, Mansson O, Butcher EC, Magnani JL. A carbohydrate domain common to both sialyl LeA and sialyl LeX is recognized by the endothelial cell leukocyte adhesion molecule ELAM-1. J Biol Chem 1991; 266:14869-14872.
66. Picker LJ, Kishimoto TK, Smith CW, Warnock RA, Butcher EC. ELAM-1 is an adhesion molecule for skin-homing T cells. Nature 1991; 349:796-799.
67. Shimizu Y, Shaw S, Graber N, Copal TV, Horgan KJ, VanSeventer GA, Newman W. Activation-independent binding of human memory T cells to adhesion molecule ELAM-1. Nature 1991; 349:799-802.
68. Lawrence MB, Springer TA. Leukocytes roll on a selectin at physiologic flow rates: distinction from and prerequisite for adhesion through integrins. Cell 1991; 65:859-873.
69. VonAndrian UH, Chambers JD, McEvoy LM, Bargatze RF, Arfors K-E, Butcher EC. Two-step model of leukocyte-endothelial cell interaction in inflammation: distinct roles for LECAM-1 and the leukocyte β2 integrins in vivo. Proc Natl Acad Sci USA 1991; 88:7538-7542.
70. Springer TA. Adhesion receptors of the immune system. Nature 1990; 346:425-434.
71. Biberfeld P, Johansson A. Contact areas of cytotoxic lymphocytes and target cells. Exp Cell Res 1975; 94:79-87.
72. Verschueren H. Interference reflection microscopy in cell biology: methodology and applications. J Cell Sci 1985; 75:279-301.
73. Bell GI, Dembo M, Bongrand P. Cell adhesion. Competition between nonspecific repulsion and specific bonding. Biophys J 1984; 45: 1051-1064.
74. Williams AF, Barclay A. In: Weir DM, Herzenberg LA, Blackwell C, Herzenberg LA, eds. Immunochemistry. Oxford: Blackwell, 1986: ch.22.
75. Kaizu T, Levery SB, Nudelman E, Stenkamp RE, Hakomori S. Novel fucolipids of human adenocarcinoma: monoclonal antibody specific for trifucosyl LeY (111^3 $FucV^3$ $FucVI^2$ $FucnLc_6$), and a possible three-dimensional epitope structure. J Biol Chem 1986; 261:11254-11258.
76. Hakomori S. Glycosphingolipids. Sci Am 1986; 254:44-53.
77. Hakomori S. Aberrant glycosylation in tumors and tumor-associated carbohydrate antigens. Adv Cancer Res 1989; 52:257-331.
78. Picker LJ, Warnock RA, Burns A, Doerschuk CM, Berg EL, Butcher EC. The neutrophil selectin LECAM-1 presents carbohydrate ligands to the vascular selectins ELAM-1 and GMP-140. Cell 1991; 66: 921-933.
79. Smith CW, Kishimoto TK, Abbass O, Hughes B, Rothlein R, McIntire LV, Butcher E, Anderson DC. Chemotactic factors regulate lectin adhesion molecule 1 (LECAM-1)-dependent neutrophil adhesion to cytokine-stimulated endothelial cells in vitro. J Clin Invest 1991;

87:609-618.
80. Kishimoto TK, Warnock RA, Jutila MA, Butcher EC, Lane C, erson DC, Smith CW. Antibodies against human neutrophil LECAM-1 (LAM-1/Leu-8/DREG-56 antigen) and endothelial cell ELAM-1 inhibit a common CD-18-independent adhesion pathway in vitro. Blood 1991; 78:805-811.
81. Arfors KE, Lundberg C, Lindbom L, Lundberg K, Beatty PG, Harlan JM. A monoclonal antibody to the membrane glycoprotein complex CD18 inhibits polymorphonuclear leukocyte accumulation and plasma leakage in vivo. Blood 1987; 69:338-340.
82. Kishimoto TK, Larson RS, Corbi AL, Dustin ML, Staunton DE, Springer TA. The leukocyte integrins. Adv Immunol 1989a; 46:149-182.
83. Carlos TM, Harlan JM. Membrane proteins involved in phagocyte adherence to endothelium. Immunol Rev 1990; 114:5-28.
84. Lawrence MB, Smith CW, Eskin SG, McIntire LV. Effect of venous shear stress on CD18-mediated neutrophil adhesion to cultured endothelium. Blood 1990; 75:227-237.
85. Kishimoto TK, Jutila MA, Berg EL, Butcher EC. Neutrophil Mac-1 and MEL-14 adhesion proteins inversely regulated by chemotactic factors. Science 1989b; 245:1238-1241.
86. Berg M, James SP. Human neutrophil release the Leu-8 lymph node homing receptor during cell activation. Blood 1990; 76:2381-2388.
87. Griffin JD, Spertini O, Ernst TJ, Belvin MP, Levine HB, Kanakura Y, Tedder TF. Granulocyte-macrophage colony-stimulating factor and other cytokines regulate surface expression of the leukocyte adhesion molecule-1 on human neutrophils, monocytes, and their precursors. J Immunol 1990; 145:576-584.
88. Singer SJ, Kupfer A. Membrane dynamics in cell-cell interactions, with application to immunology. In: Davis MM, Kappler J, eds. The T-Cell Receptor. New York: Alan R. Liss, 1988:361-376.
89. Singer SJ. Intercellular communication and cell-cell adhesion. Science 1992; 255:1671-1677.
90. Singer SJ. Fluid mosaic model of membrane structure: some applications to ligand-receptor and cell-cell interactions. In: Bradshaw RA, Frazier WA, Merrell RC, Gottlieb DI, Hogue-Angeletti RA, eds. Surface Membrane Receptors, Interface Between Cells and Their Environment. New York: Plenum Press, 1976:1-24.
91. Weis RM, Balakrishnan K, Smith BA, McConnell HM. Stimulation of fluorescence in a small contact region between rat basophil leukemia cells and a planar lipid membrane targets by coherent evanescent radiation. J Biol Chem 1982; 257:6440-6445.
92. McCloskey MA, Contact-induced redistribution of specific membrane components: local accumulation and development of adhesion. J Cell Biol 1986; 102:2185-2196.
93. Hannun YA, Bell RM. Lysosphingolipids inhibit protein kinase C:

Implication for the sphingolipidoses. Science 1987; 235:670-674.
94. Hannun YA, Bell RM. Functions of sphingolipids and sphingolipid breakdown products in cellular regulation. Science 1989; 243:500-507.
95. Zhang H, Buckely NE, Gibson K, Spiegel S. Sphingosine stimulates cellular proliferation via protein kinase C-independent pathway. J Biol Chem 1990; 265:76-81.
96. Weis FMB, Davis RJ. Regulation of epidermal growth factor receptor signal transduction: Role of gangliosides. J Biol Chem 1990; 265:12059-12066.
97. Hanai N, Dohi T, Nores GA, Hakomori S. A novel ganglioside, de-N-acetyl-GM3 (II^3NeuNH_2Lac Cer), acting as a strong promoter for epidermal growth factor receptor kinase and as a stimulator for cell growth. J Biol Chem 1988; 263:6296-6301.
98. Igarashi Y, Nojiri H, Hanai N, Hakomori S. Gangliosides that modulate membrane protein function. Meth Enzymol 1989; 179:521-541.
99. Nishizuka Y. The role of protein kinase C in cell surface signal transduction and tumor promotion. Nature 1984; 308:693-698.
100. Berridge MJ. Rapid accumulation of inositol triphosphate reveals that agonists hydrolyse polyphosphoinositides instead of phosphatidylinositol. Biochem J 1983; 212:849-858.
101. Exton JH. Singaling through phosphatidylcholine breakdown. J Biol Chem 1990; 265:1-4.
102. Elices MJ, Urry LA, Hemler ME. Receptor functions for the integrin VLA-3: fibronectin, collagen, and laminin binding are differentially influenced by Arg-Gly-Asp peptide and by divalent cations. J Cell Biol 1991; 112:169-181.
103. Wayner EA, Orlando RA, Cheresh DA. Integrins $\alpha_v\beta_3$ and $\alpha_v\beta_5$ contribute to cell attachment to vitronectin but differentially distribute on the cell surface. J Cell Biol 1991; 113:919-929.
104. Chan BMC, Kassner PD, Schiro JA, Byers R, Kupper TS, Hemler ME. Distinct cellular functions mediated by different VLA integrin α subunit cytoplasmic domains. Cell 1992; 68:1051-1060.
105. Stepp MA, Spurr-Michaud S, Tisdale A, Elwell J, Gipson IK. $\alpha_6\beta_4$ integrin heterodimer is a component of hemidesmosomes. Proc Natl Acad Sci USA 1990; 87:8970-8974.
106. Sonnenberg A, Calafat J, Janssen H, Daams H, van der Raaij-Helmer LM, Falcioni R, Kennel SJ, Aplin JD, Baker J, Loizidou M, Garrod D. Integrin $\alpha_{6/}\beta_4$ complex is located in hemidesmosomes, suggesting a major role in epidermal cell-basement membrane adhesion. J Cell Biol 1991; 113:907-917.
107. Kurpakus MA, Quaranta V, Jones JCR. Surface relocation of $\alpha_6\beta_4$ integrins and assembly of hemidesmosomes in an in vitro model of wound healing. J Cell Biol 1991; 115:1737-1750.
108. Elices MJ, Hemler ME. The human integrin VLA-2 is a collagen receptor on some cells and a collagen/laminin receptor on others. Proc Natl Acad Sci USA 1989; 86:9906-9910.
109. Kirchhofer D, Languino LR, Ruoslahti E, Pierschbacher MD. $\alpha_2\beta_1$

integrins from different cell types show different binding specificities. J Biol Chem 1990b; 265:615-618.
110. Staatz WD, Rajpara SM, Wayner EA, Carter WG, Santoro SA. The membrane glycoprotein la-lla (VLA-2) complex mediates the Mg^{++}-dependent adhesion of platelets to collagen. J Cell Biol 1989; 108:1917-1924.
111. Wright SD, Silverstein SC. Tumor-promoting phorbol esters stimulate C3b and C3bi receptor-mediated phagocytosis in cultured human monocytes. J Exp Med 1982; 156:1149-1164.
112. Wright SD, Rao PE, Van Voorhis WC, Craigmyl LS, Jida K, Talle MA, Westberg EF, Goldstein G, Silverstein SC. Identification of the C3bi receptor on human monocytes and macrophages by using monoclonal antibodies. Proc Natl Acad Sci USA 1983; 80:5699-5703.
113. Wright SD, Licht MR, Craigmyl LS, Silverstein SC. Communication between receptors for different ligands on a single cell: ligation of fibronectin receptors induces a reversible alteration in the function of complement receptors on cultured human monocytes. J Cell Biol 1984; 99:336-339.
114. Dustin ML, Springer TA. Role of lymphocyte adhesion receptors in transient interactions and cell locomotion. Annu Rev Immunol 1991; 9:2-66.
115. van Kooyk Y, van de Wiel-van Kamenade P, Weder P, Kuijpers TW, Figdor CG. Enhancement of LFA-1-mediated cell adhesion by triggering through CD2 or CD3 on T lymphocytes. Nature 1989; 342:811-813.
116. Hemler ME. VLA proteins in the integrin family: structures, functions and their role on leukocytes. Annu Rev Immunol 1990; 8:365-400.
117. Shimizu Y, Shaw S. Lymphocyte interactions with extracellular matrix. FASEB J 1991; 5:2292-2299.
118. Ferguson TA, Mizutani H, Kupper TS. Two integrin-binding peptides abrogate T cell-mediated immune responses in vivo. Proc Natl Acad Sci USA 1991; 88:8072-8076.
119. Shimizu Y, van Seventer G, Horgan KJ, Shaw S. Regulated expression and binding of three VLA ($\beta 1$) integrin receptors on T cells. Nature 1990; 345:250-253.
120. Chan BM, Wong JG, Rao A, Hemler ME. T cell receptor-dependent, antigen-specific stimulation of a murine T cell clone induces a transient, VLA protein-mediated binding to extracellular matrix. J Immunol 1991; 147:398-404.
121. Wilkins JA, Stupack D, Stewart S, Caixia S. β_1 integrin-mediated lymphocyte adherence to extracellular matrix is enhanced by phorbol ester treatment. Eur J Immunol 1991; 21:517-522.
122. Tapley P, Horwitz AF, Buck CA, Burridge K, Duggan K, Hirst R, Rohrschneider L. Analysis of the avain fibronectin receptor (integrin) as direct substrate for $pp60^{v-src}$. Oncogene 1989; 4:325-333.
123. Horvath AR, Elmore MA, Kellie S. Differential tyrosine-specific phos-

phorylation in integrin in Rous sarcoma virus transformed cells with differing transformed phenotypes. Oncogene 1990; 5:1349-1357.
124. Guan JL, Trevithick JE, Hynes RO. Fibronectin/integrin interaction induces tyrosine phosphorylation of a 120kDa protein. Cell Regul 1991; 2:951-964.
125. Schwartz MA, Both G, Lechene C. Effect of cell spreading on cytoplasmic pH in normal and transformed fibroblasts. Proc Natl Acad Sci USA 1989; 86:4525-4529.
126. Schwartz MA, Ingber DE, Lawrence M, Springer TA, Lechene C. Multiple integrins share the ability to induce elevation of intracellular pH. Exp Cell Res 1991; 195:533-535.
127. Ingber DE, Prusty D, Frangioni JV, Cragoe EJ Jr., Lechene C, Schwartz MA. Control of intracellular pH and growth by fibronectin in capillary endothelia cells. J Cell Biol 1990; 110:1803-1811.
128. Schwartz MA, Rupp EE, Frangioni JV, Lechene CP. Cytoplasmic pH and anchorage-independent growth induced by v-Ki-ras, v-src or polyoma middle T. Oncogene 1990; 5:55-58.
129. Obara M, Kang MS, Yamada KM. Site-directed mutagenesis of the cell-binding domain of human fibronectin: separable, synergistic sites mediate adhesive function. Cell 1988; 53:649-657.
130. Woods A, Johansson S, Hook M. Fibronectin fibril formation involves cell interactions with two fibronectin domains. Exp Cell Res 1988; 177:272-283.
131. Albelda SM, Mette SA, Elder DE, Stewart R, Damjanovich K, Herlyn M, Buck CA. Integrin distribution in malignant melanomas: association of the β3 subunit with tumor progression. Cancer Res 1990; 50:6757-6764.
132. Dedhar S, Saulnier R. Alterations in integrin receptor expression on chemically transformed human cells: Specific enhancement of laminin and collagen receptor complexes. J Cell Biol 1990; 110:481-489.
133. Chan BM, Matsuura N, Takada Y, Zetter BR, Hemler ME. In vitro and in vivo consequences of VLA-2 expression on rhabdomyosarcoma cells. Science 1991; 251:1600-1602.
134. Humphries MJ, Olden K, Yamada KM. A synthetic peptide from fibronectin inhibits experimental metastasis of murine melanoma cells. Science 1986; 233:467-470.
135. Iwamoto Y, Robey FA, Graf J, Sasaki M, Kleinman HK, Yamada Y, and Martin GR. YIGSR, a synthetic laminin pentapeptide inhibits experimental metastasis formation. Science 1987; 238:1132-1134.
136. Saiki I, Iida J, Murata J, Ogawa R, Nishi N, Sugimura K, Tokura S, Azuma I. Inhibition of the metastasis of murine malignant melanoma by synthetic polymeric peptides containing core sequences of cell-adhesive molecules. Cancer Res 1989; 49:3815-3822.
137. Ugen KE, Mahalingam M, Klein PA, Kao, K.-J. Inhinition of tumor cell-induced platelet aggregation and the experimental tumor metastasis by the synthetic Gly-Arg-Gly-Asp-Ser peptide. J Natl Can-

cer Inst 1988; 80:1461-1466.
138. Gehlsen KR, Argraves WS, Pierschbacher MD, Ruoslahti E. Inhibition of in vitrto tumor cell invasion by Arg-Gly-Asp-containing synthetic peptides. J Cell Biol 1988; 106:925-930.
139. Glinsky GV. Glycoamines: Biochemistry of a new class of humoral tumor markers. J Tumor Marker Oncology 1989; 4:193-221.
140. Glinsky GV, Surgova TM, Sidorenko MV, Vinnitsky VB, Kolesnik LA, Shvachko LA. Interaction of glycoamines and polyamines within the system of α_2 macroglobulin-dependent regulation of plasma transglutaminase. A possible molecular mechanism of invasion and implantation in embryogenesis and cancer. J Tumor Marker Oncology 1990; 5:161-166.
141. Yamada KM, Yamada SS, Pastan I. Cell surface protein partially restores morphology, adhesiveness, and contact inhibition of movement to transformed fibroblasts. Proc Natl Acad Sci USA 1976; 73:1217-1221.
142. Hayman EG, Engwall E, Ruoslahti E. Concomitant loss of cell surface fibronectin and laminin from transformed rat kidney cells. J Cell Biol 1981; 88:352-357.
143. Ali IU, Mautner VM, Lanza RP, Hynes RO. Restoration of normal morphology, adhesion and cytoskeleton in transformed cells by addition of a transformation-sensitive surface protein. Cell 1977; 11:115-126.
144. Loewenstein WR. Junctional intercellular communication and the control of growth. Biochim Biophys Acta 1979; 560:1-65.
145. Suzuki S, Sano K, Tanihara H. Diversity of the cadherin family: evidence for eight new cadherins in nervous tissue. Cell Regulat 1991; 2:261-270.
146. Miyatani S, Shimamura K, Hatta M, Nagafuchi A, Nose A, Matsunaga M, Hatta K, Takeichi M. Science 1989; 245:631-635.
147. Friedlander DR, Mege R-M, Cunningham BA, Edelman GM. Cell sorting-out is modulated by both the specificity and amount of different cell adhesion molecules (CAMs) expressed on cell surfaces. Proc Natl Acad Sci USA 1989; 86:7043-7047.
148. Nose A, Tsuji K, Takeichi M. Localization of specificity determining sites in cadherin cell adhesion molecules. Cell 1990; 61:147-155.
149. Nagafuchi A, Takeichi M. Cell binding function of E-cadherin is regulated by the cytoplasmic domain. EMBO J 1988; 7:2679-2684.
150. Nagafuchi A, Takeichi M. Transmembrane control of cadherin-mediated cell adhesion: A 94 kDa protein functionally associated with a specific region of the cytoplasmic domain of E-cadherin. Cell Regulat 1990; 1:37-45.
151. Ozawa M, Baribault H, Kemler R. The cytoplasmic domain of the cell adhesion molecule uvomorulin associates with three independent proteins structurally related in different species. EMBO J 1989; 8:1711-1717.
152. Ozawa M, Ringwald M, Kemler R. Uvomorulin-catenin complex for-

mation is regulated by a specific domain in the cytoplasmic region of the cell adhesion molecules. Proc Natl Acad Sci USA 1990; 87:4246-4250.
153. Fujimori T, Miyatani S, Takeichi M. Ectopic expression of N-cadherin perturbs histogenesis of *Xenopus* embryos. Development 1990; 110:97-93.
154. Hirano S, Nose A, Hatta K, Kawakami A, Takeichi M. Calcium-dependent cell-cell adhesion molecules (cadherins): subclass specificities and possible involvement of actin bundles. J Cell Biol 1987; 105:2501-2510.
155. Nelson WJ, Shore EM, Wang AZ, Hammerton RW. Identification of a membrane-cytoskeletal complex containing the cell adhesion molecule uvomovulin (E-cadherin), ankyrin, and fodrin in Madin-Darby canine kidney epithelial cells. J Cell Biol 1990; 110:349-357.
156. Matsuzaki F, Mege R-M, Jaffe SH, Friedlander DR, Gallin WJ, Goldberg JI, Cunningham BA, Edelman GM. cDNAs of cell adhesion molecules of different specificity induce changes in cell shape and border formation in cultured S180 cells. J Cell Biol 1990; 110:1239-1252.
157. Geiger B. Cytoskeleton-associated cell contacts. Curr Opin Cell Biol 1989; 1:103-107.
158. Rodriguez-Boulan E, Nelson WJ. Morphogenesis of the polarized epithelial cell phenotype. Science 1989; 245:718-725.
159. Kanno Y, Sasaki Y, Shiba Y, Yoshida-Noro C, Takeichi M. Monoclonal antibody ECCD-1 inhibits intercellular commuinication in teratocarcioma PCC3 cells. Exp Cell Res 1984; 152:270-274.
160. Gumbiner B, Simons K. A functional assay for proteins involved in establishing an epithelial occluding barrier: identification of a uvomorulin-like polypeptide. J Cell Biol 1986; 102:457-468.
161. Gumbiner B, Stevenson B, Grimaldi A. The role of the cell adhesion molecule uvomorulin in the formation and maintenance of the epithelial junctional complex. J Cell Biol 1988; 107:1575-1587.
162. Mege RM, Matsuzaki F, Gallin WJ, Goldberg JI, Cunningham BA, Edelman GM. Construction of epithelioid sheets by transfection of mouse sarcoma cells with cDNAs for chicken cell adhesion molecules. Proc Natl Acad Sci USA 1988; 85:7274-7278.
163. Hashimoto M, Niwa O, Nitta Y, Takeichi M, Yokoro K. Unstable expression of E-cadherin adhesion molecules in metastatic ovarian tumor cells. Jpn J Cancer Res 1989; 80:459-465.
164. Glinsky GV. Cell adhesion and metastasis: is the site specificity of cancer metastasis determined by leukocyte-endothelial cell recognition and adhesion? Crit Rev Onc/Hemat 1993; 14:229-278.
165. Fukuda M. Possible role of tumor-associated carbohydrate antigens. Cancer Res 1996; 56:2237-2244.
166. Hakomori S. Tumor malignancy defined by aberrant glycosylation and sphingo(glyco)lipid metabolism. Cancer Res 1996; 56:5309-5318.
167. Springer TA. Traffic signals for lymphocyte recirculation and leu-

kocyte emigration: the multistep paradigm. Cell 1994; 76:301-314.
168. Ruoslahti E, Reed JC. Anchorage dependence, integrins, and apoptosis. Cell 1994; 77:477-478.
169. Pienta KJ, Naik H, Akhtar A, Yamazaki K, Replogle TS, Lehr J, Donat TL, Tait L, Hogan V, Raz A. Inhibition of spontaneous metastasis in rat prostate cancer model by oral administration of modified citrus pectin. J Natl Cancer Inst 1995; 87:348-353.
170. Glinsky GV, Price JE, Glinsky VV, Mossine VV, Kiriakova G, Metcalf JB. Inhibition of human breast cancer metastasis in nude mice by synthetic glycoamines. Cancer Res 1996; 56:5319-5324.
171. Glinsky GV, Mossine VV, Price JE, Bielenberg D, Glinsky VV, Ananthaswamy HN, Feather MS. Inhibition of colony formation in agarose of metastatic human breast carcinoma and melanoma cells by synthetic glycoamine analogs. Clin Exp Metastasis 1996; 14:253-267.
172. Peletskaya EN, Glinsky GV, Deutscher SL, Quinn TP. Identification of peptide sequences that bind the Thomsen-Friedenreich cancer-associated glycoantigen from bacteriophage peptide display libraries. Mol Diversity 1996; 2:13-18.
173. Peletskaya EN, Glinsky VV, Glinsky GV, Deutscher SL, Quinn TP. Characterization of peptides that bind the tumor-associated Thomsen-Friedenreich antigen selected from bacteriophage display libraries. J Mol Biol 1977; 270:374-384.
174. Glinsky GV, Glinsky VV. Apoptosis and metastasis: a superior resistance of metastatic cancer cells to the programmed cell death. Cancer Lett 1996; 101:43-51.
175. Glinsky GV, Glinsky VV, Ivanova AB, Hueser CN. Apoptosis and metastasis: increase apoptosis resistance of metastatic cancer cells is associated with the profound deficiency of apoptosis execution mechanisms. Cancer Lett 1997; 115:185-193.
176. Glinsky GV. Apoptosis in metastatic cancer cells. Crit Rev Oncol/Hemat 1997; 25:175-186.
177. Liotta LA. Tumor invasion and metastases—role of the extracellular matrix. Cancer Res 1986; 46:1-7.
178. Roos E. Cellular adhesion, invasion and metastasis. Biochim Biophys Acta 1984; 738:263-284.
179. Mullins DE, Rorlich ST. The role of proteinases in cellular invasiveness. Biochim Biophys Acta 1983; 695:177-214.
180. Ossowski L. Plasminogen activator dependent pathways in the dissemination of human tumor cells in the chick embryo. Cell 1988; 52:321-328.
181. Axelrod JH, Reich R, Miskin R. Expression of human recombinant plasminogen activators enhances invasion and experimental metastasis of H-ras-transformed NIH 3T3 cells. Mol Cell Biol 1989; 9:2133-2141.
182. Mueller-Kleiser W. Multicellular spheroids. A review on cellular

aggregates in cancer research. J Cancer Res Clin Oncol 1987; 113:101-122.
183. Sutherland RM. Cell and environment interactions in tumor microregions: The multicell spheriod model. Science 1988; 240:177-184.
184. Butcher EC. The regulation of lymphocyte traffic. Curr Topics Microbiol Immun 1986; 128:85-122.
185. Yednock TA, Rosen SD. Lymphocyte homing. Adv Immun 1989; 44:313-378.
186. Stoolman LM. Adhesion molecules controlling lymphocyte migration. Cell 1989; 56:907-910.
187. Springer TA, Lasky LA. Sticky sugars for selectins. Nature 1991; 349:196-197.
188. Glinsky GV. The blood group-related glycoepitopes: key structural determinants in immunogenesis and AIDS pathogenesis. Medical Hypotheses 1992; 39:212-224.
189. Hart GW. The role of asparagine-linked oligosaccharide in cellular recognition by thymic lymphocytes. J Biol Chem 1982; 257:151-158.
190. Cullen SE, Kindle CS, Shreffler DC, Cowing C. Differential glycosylation of murine B cell and spleen adherent cell Ia antigens. J Immunol 1981; 127:1478-1484.
191. Cowing C, Chapdelaine JM. T cells discriminate between Ia antigens expressed on allogeneic accessory cells and B cells: A potential function for carbohydrate side chains on Ia molecules. Proc Natl Acad Sci USA 1983; 80:6000-6004.
192. Taiara S, Kakiuchi T, Minotti M, Nariuchi H. The regulatory role of sialic acids in the response of class II reactive T cell hybridomas to allogenic B cells. J Immunol 1986; 137:2448-2454.
193. Powell LD, Whiteheart SW, Hart GW. Cell surface sialic acid influences tumor cell recognition in the mixed lymphocyte reaction. J Immunol 1987; 139:262-270.
194. Black P, Vitetta E, Forman J, Kang C-Y, May R, Uhr J. Role of glycosylation in the H-2 restricted cytolysis of virus-infected cells. Eur J Immunol 1981; 11:48-55.
195. Booy CJP, Needjes LJ, Boes J, Ploegh HL, Melief CJM. Specific immune responses restored by alteration in carbohydrate chains of surface molecules on antigen-presenting cells. Eur J Immunol 1989; 19:537-542.
196. Adachi M, Hayami M, Kashiwagi N, Mizuta T, Ohta Y, Gill MJ, Matheson DS, Tamaoki T, Shiozawa C, Hakomori S. Expression of LeY antigen in human immunodeficiency virus-infected human T cell lines and in peripheral lymphocytes of patients with acquired immune deficiency syndrome (AIDS) and AIDS-related complex (ARC). J Exp Med 1988; 167:323-331.
197. Andrews PW, Gonczol E, Fenderson BA, Holmes EH, O'Malley G, Hakomori S, Plotkin S. Human cytomegalovirus induces stage-specific embryonic antigen 1 in differentiating human teratocarcinoma cells and fibroblasts. J Exp Med 1989; 169:1347-1359.
198. Lloyd KO, Old LJ. Human monoclonal antibodies to glycolipids and

other carbohydrate antigens: dissection of the humoral immune response in cancer patients. Cancer Res 1989; 49:3445-3451.
199. Huesmann M, Scott B, Kisielow P, von Boehmer H. Kinetics and efficacy of positive selection in the thymus of normal and T cell receptor transgenic mice. Cell 1991; 66:533-540.
200. Despont JP, Abel CA, Grey HM. Sialic acids and sialyltransferases in murine lymphoid cells: indicators of T cell maturation. Cell Immun 1975; 17:487-494.
201. Shortman K, von Boehmer H, Lipp J, Hopper K. Subpopulations of T-lymphocytes. Transplant Rev 1975; 25:163-210.
202. Butcher EC, Rouse RV, Coffman RL, Nottenburg CN, Hardy RR, Weissman IL. Surface phenotype of Peyer's patch germinal center cells: implications for the role of germinal centers in B cell differentiation. J Immun 1982; 129:2698-2707.
203. Diamond MS, Staunton DE, Marlin SD, Springer TA. Binding of the integrin Mac-1 (CDIIb/CD18) to the third immunoglobulin-like domain of ICAM-1 (CD54) and its regulation by glycosylation. Cell 1991; 65:961-971.
204. Mackay CR, Martson WL, Dudler L. Naive and memory T cells show distinct pathways of lymphocyte recirculation. J Exp Med 1990; 171:801-817.
205. Bevilacqua MP, Pober JS, Mendrick DL, Cotran RS, Gimbrone MA. Identification of an inducible endothelial-leukocyte adhesion molecule. Proc Natl Acad Sci USA 1987; 84:9238-9242.
206. Lo SK, Detmers PA, Levin SM, Wright SD. Transient adhesion of neutrophils to endothelium. J Exp Med 1989; 169:1779-1793.
207. Hakomori SI. Aberrant glycosylation in cancer cell membranes as focused on glycolipids: overview and perspectives. Cancer Res 1985; 45:2405-2414.
208. Feizi T. Demonstration by monoclonal antibodies that carbohydrate structures of glycoproteins and glycolipids are onco-developmental antigens. Nature 1985; 314:53-57.
209. Hakomori S, Kannagi R. Glycosphingolipids as tumor-associated and differentiation markers. J Natl Cancer Inst 1983; 71:231-251.
210. Miyake M, Zenita K, Tanaka O, Okada Y, Kannagi R. Stage-specific expression of SSEA-1-related antigens in the developing lung of human embryos and its relation to the distribution of these antigens in lung cancers. Cancer Res 1988; 48:7150-7158.
211. Solter D, Knowles BB. Monoclonal antibody defining a stage-specific mouse embryonic antigen (SSEA-1). Proc Natl Acad Sci USA 1978; 75:5565-5569.
212. Gooi HC, Feizi T, Kapadia A, Knowles BB, Solter D, Evans MJ. Stage-specific embryonic antigen involves α(1-3) fucosylated type 2 blood group chains. Nature 1981; 292:156-158.
213. Kannagi R, Nudelman E, Levery SB, Hakomori S. A series of human erythrocyte glycosphingolipids reacting to the monoclonal antibody directed to a developmentally regulated antigen SSEA-1. J Biol

Chem 1982; 257:14865-14874.
214. Abe K, McKibbin JM, Hakomori SI. The monoclonal antibody directed to difucosylated type 2 chain. J Biol Chem 1983; 258: 11793-11797.
215. Fukushi Y, Kannagi R, Hakomori S, Shepard T, Kulander BG, Singer JW. Localization and distribution of difucoganglioside ($VI^3NueAcV^3III^3Fuc_2Lc_6$) in normal and tumor tissues defined by its monoclonal antibody FH6. Cancer Res 1985; 45:3711-3717.
216. Hakomori S, Nudelman E, Levery SB, Kannagi R. Novel fucolipids accumulating in human adenocarcinoma I. Glycolipids with di- or trifucosylated type 2 chain. J Biol Chem 1984; 259:4681-4685.
217. Kannagi R, Fukushi Y, Tachikawa T, Noda A, Shin S, Shigeta K, Hiraiwa N, Fukuda Y, Inamoto T, Hakomori S, Imura H. Quantitative and qualitative characterization of human cancer-associated serum glycoprotein antigens expressing fucosyl or sialyl-fucosyl type 2 chain polylactosamine. Cancer Res 1986; 46:2619-2626.
218. Shi ZR, McIntyre LJ, Knowles BB, Solter D, Kim YS. Expression of a carbohydrate differentiation antigen, stage-specific embryonic antigen 1, in human colonic adenocarcinoma. Cancer Res 1984; 44:1142-1147.
219. Spitalnik SL, Spitalnik PF, Dubois C, Mulshine J, Magnani JL. Cuttitta F, Civin CI, Minna JD, Ginsburg V. Glycolipid antigen expression in human lung cancer. Cancer Res 1986; 46:4751-4755.
220. Cuttitta F, Rosen S, Gazdar AF, Minna JD. Monoclonal antibodies that demonstrate specificity for several types of human lung cancer. Proc Natl Acad Sci USA 1981; 78:4591-4595.
221. Huang LC, Brockhaus M, Magnani JL, Cuttitta F, Rosen S, Minna J, Ginsburg V. Many monoclonal antibodies with an apparent specificity for certain lung cancers are directed against a sugar sequence found in lacto-N-fucopentaose III. Arch Biochem Biophys 1983; 220:318-320.
222. Chia D, Terasaki PI, Suyama N, Galton J, Hirota M, Katz D. Use of monoclonal antibodies to sialylated Lewis-x and sialylated Lewis-a for serological tests of cancer. Cancer Res 1985; 45:435-437.
223. Zenita K, Kirihata Y, Kitahara A, Shigeta K, Higuchi K, Hirashima K, Murachi T, Miyake M, Takeda T, Kannagi R. Fucosylated type-2 chain polylactosamine antigens in human lung cancer. Int J Cancer 1988; 41:344-349.
224. Itzkowitz SH, Shi ZR, Kim YS. Heterogenous expression of two oncodevelopmental antigens, CEA and SSEA-1, in colorectal cancer. Histochem J 1986; 18:155-163.
225. Andrews PW, Goodfellow PN, Shevinsky LH, Bronson DL, Knowles BB. Cell-surface antigens of a clonal human embryonal carcinoma cell line: morphological and antigenic differentiation in culture. Int J Cancer 1982; 29:523-531.
226. Combs SG, Marder RJ, Minna JD, Mulshine JL, Polovina MR, Rosen

ST. Immunohistochemical localization of the immunodominant differentiation antigen lacto-N-fucopentaose III in normal adult and fetal tissues. J Histochem Cytochem 1984; 32:982-988.
227. Fox N, Damjanov I, Martinez-Hernandez A, Knowles BB, Solter D. Immunohistochemical localization of the early embryonic antigen (SSEA-1) in postimplantation mouse embryos and fetal and adult tissues. Dev Biol 1981; 83:391-398.
228. Fox N, Damjanov I, Knowles BB, Solter D. Immunohistochemical localization of the mouse stage-specific embryonic antigen 1 in human tissues and tumors. Cancer Res 1983; 43:669-678.
229. Yamamoto M, Boyer AM, Schwarting GA. Fucose-containing glycolipids are stage-and region-specific antigens in developing embryonic brain of rodents. Proc Natl Acad Sci USA 1985; 82:3045-3049.
230. Fenderson BA, Zehavi U, Hakomori S. A multivalent lacto-N-fucopentaose III—lysyllysine conjugate decompacts preimplantation mouse embryos, while the free oligosacharide is ineffective. J Exp Med 1984; 160:1591-1596.
231. Shur BD. Embryonal carcinoma cell adhesion: The role of surface galactosyltransferase and its 90K lactosaminoglycan substrate. Dev Biol 1983; 99:360-372.
232. Takeichi M. Cadherins: A molecular family essential for selective cell-cell adhesion and animal morphogenesis. Trends Gen 1987; 3:213-217.
233. Bayna EA, Shaper JH, Shur BD. Temporally specific involvement of cell surface β1, 4 galactosyltransferase during mouse embryo morula compaction. Cell 1988; 53:145-157.
234. Lee S, Gilula NB, Warner AE. Gap junctional communication and compaction during preimplantation stages of mouse development. Cell 1987; 51:851-860.
235. Fidler IJ, Gersten DM, Hart IR. The biology of cancer invasion and metastasis. Adv Cancer Res 1978; 28:149-250.
236. Schirrmacher V. Cancer metastasis: experimental approaches, theoretical concepts, and impacts for treatment strategies. Adv Cancer Res 1985; 43:1-73.
237. Dano K, Reasen PA, Grondahl-Hansen J, Kristensen BP, Nielsen LS, Skriver L. Plasminogen activatiors, tissue degradation and cancer. Adv Cancer Res 1985; 44:139-266.
238. Reich E. Activation of plasminogen: a general mechanism for producing localized extracellular proteolysis. In: Berlin RD, Herrmann H, Lepow IH, Tanzer JM, eds. Molecular Basis of Biological Degradation Processes. New York: Academic Press, 1978:155-169.
239. Reich R, Thompson E, Iwamoto Y, Martin GR, Deason JR, Fuller GC, Miskin R. Effects of inhibitors of plasminogen activator, serine proteinases and collagenase IV on the invasion of basement membrane by metastatic cells in mice and humans. Cancer Res 1987; 48:3307-3312.
240. Mignotti P, Robbins E, Rifkin DB. Tumor invasion through the hu-

man amniotic membrane: requirement for a proteinase cascade. Cell 1986; 47:487-498.
241. Ossowski L, Reich E. Antibodies to plasminogen activator inhibit human tumor metastasis. Cell 1983; 35:611-619.
242. Skriver L, Larsson LI, Keilberg V, Nielsene LS, Resen PB, Kristensen P, Dano K. Immunocytochemical localization of urokinase-type plasminogen activator in Lewis lung carcinoma. J Cell Biol 1984; 99:753-757.
243. Hearing VJ, Law LW, Corti A, Appella E, Blasi F. Modulation of metastatic potential by cell surface urokinase of murine melanoma cells. Cancer Res 1988; 48:1270-1278.
244. Mareel MM, VanRoy FM, DeBaetselier P. The invasive phenotypes. Cancer Metastasis Rev 1990; 9:45-62.
245. Zeidman I. The fate of circulating tumor cells. I. Passage of cells through capillaries. Cancer Res 1961; 21:38-39.
246. Fisher B, Fisher ER. Transmigration of lymph nodes by tumor cells. Science 1966; 152:1397-1398.
247. Kurokawa Y. Experiments on lymph node metastasis by intralymphatic inoculation of rat ascites tumor cells, with special reference to lodgement, passage and growth of tumor cells in lymph nodes. Gann 1970; 61:461-470.
248. Nicolson GL. Cancer metastasis. Organ colonization and the cell-surface properties of malignant cells. Biochim Biophys Acta 1982; 695:113-176.
249. Nicolson GL. Cell surface molecules and tumor metastasis. Regulation of metastatic phenotypic diversity. Exp Cell Res 1984; 150:3-22.
250. Glaves D. Correlation between circulating cancer cells and incidence of metastases. Br J Cancer 1983; 48:665-673.
251. Liotta LA, Kleinerman J, Saidel GM. The significance of hematogenous tumor cell clumps in the metastatic process. Cancer Res 1976; 36:889-894.
252. Varani J, Orr W, Ward PA. Adhesive characteristics of tumor cell variant of high and low tumorigenic potential. J Natl Cancer Inst 1980; 64:1173-1178.
253. Winkelhake JL, Nicolson GL. Determination of adhesive properties of variant metastatic melanoma cells to BALB/3T3 cells and their virus-transformed derivatives by a monolayer attachment assays. J Natl Cancer Inst 1976; 56:285-291.
254. Fidler IJ. Selection of successive tumor lines for metastasis. Nat New Biol 1973; 242:148-149.
255. Nicolson GL, Brunson KW, Fidler IJ. Specificity of arrest, survival, and growth of selected metastatic variant cell lines. Cancer Res 1978; 38:4105-4111.
256. Fidler IJ, Bucana C. Mechanism of tumor cell resistance to lysis by syngeneic lymphocytes. Cancer Res 1977; 37:3945-3956.
257. Gasic GL, Gasic TB, Galanti N, Johnson T, Murphy S. Platelet-tu-

mor cell interactions in mice. The role of platelets in the spread of malignant disease. Int J Cancer 1973; 11:704-718.
258. Nicolson GL, Winkelhake JL. Organ specificity of blood-borne tumor metastasis determined by cell adhesion? Nature (Lond) 1975; 255:230-232.
259. Raz A, Bucana C, McLellan W, Fidler IJ. Distribution of membrane anionic sites on B16 melanoma variants with differing lung colonising potential. Nature (Lond) 1980; 284:363-364.
260. Hausman RE. Increase in homotypic aggregation of metastatic Morris hepatoma cells after fusion with membranes from non-metastatic cells. Int J Cancer 1983; 32:603-608.
261. Oz OK, Campbell A, Tao T. Reduced cell adhesion to fibronectin and laminin is associated with altered glycosylation of β1 integrins in a weakly metastatic glycosylation mutant. Int J Cancer 1989; 44:343-347.
262. Tao TW, Johnson LK. Altered adhesiveness of tumor cell surface variants with reduced metastasizing capacity. Reduced adhesiveness to vascular wall components in culture. Int J Cancer 1982; 30:763-766.
263. Tao TW, Jenkins JM, Vosbeck K, Matter A, Miller M, Jockusch BM, Schen Z, Burger MM. Lectin-resistant variants of mouse melanoma cells. II. In vitro characteristics. Int J Cancer 1983; 31:239-247.
264. Damsky CH, Knudsen KA, Buck CA. Integral membrane glycoproteins in cell-cell and cell-substratum adhesion. In: Ivatt R, ed. Biology of Glycoproteins. New York: Academic Press. 1983:1-64.
265. Lotan R, Raz A. Low colony formation in vivo and in culture as exhibited by metastatic melanoma cells selected for reduced homotypic aggregation. Cancer Res 1983; 43:2088-2093.
266. Lotan R, Lotan D, Raz A. Inhibition of tumor cell colony formation in culture by a monoclonal antibody to endogenous lectins. Cancer Res 1985; 45:4349-4353.
267. Lotan R, Cavvalero D, Lotan D, Raz A. Biochemical and immunological characterization of K-1735P melanoma galactoside-binding lectins and their modulation by differentiation inducers. Cancer Res 1989; 49:1261-1268.
268. Meromsky L, Lotan R, Raz A. Implications of endogenous tumor cell surface lectins as mediators of cellular interactions and lung colonization. Cancer Res 1986; 46:5270-5275.
269. Raz A, Lotan R. Lectin-like activities associated with human and murine neoplastic cells. Cancer Res 1981; 41:3642-3647.
270. Barondes SH. Soluble lectins: a new class of extracellular proteins. Science 1984; 223:1259-1264.
271. Briles EB, Gregory W, Fletcher P, Kornfeld S. Vertebrate lectins, Comparison of properties of beta-galactoside-binding lectins from tissues of calf and chicken. J Cell Biol 1979; 81:528-537.
272. Cerra RF, Haywood-Reid PL, Barondes SH. Endogenous mamma-

lian lectin localized extracellularly in lung elastic fibers. J Cell Biol 1984; 98:1580-1589.
273. Nowak TP, Kobiler D, Roel LE, Barondes SH. Developmentally regulated lectin from embryonic chick pectoral muscle. Purification by affinity chromatography. J Biol Chem 1977; 252:6026-6030.
274. Zalik SE, Thompson LW, Ledsham IM. Expression of an endogenous galactose-binding lectin in the early chick embryo. J Cell Sci 1987; 88:483-493.
275. Cook GM, Zalik SE, Milos N, Scott VA. A lectin which binds specifically to beta-D-galactoside groups is present at the earliest stages of chick embryo development. J Cell Sci 1979; 38:293-304.
276. Drickamer K. Two distinct classes of carbohydrate-recognition domains in animal lectins. J Biol Chem 1988; 263:9557-9560.
277. Harrison FL, Chesterton CJ. Factors mediating cell-cell recognition and adhesion. Galaptins, a recently discovered class of bridging molecules. FEBS Lett 1980; 122:157-165.
278. Monsigny M, Kieda C, Roche AC. Membrane glycoproteins, glycolipids and membrane lectins as recognition signals in normal and malignant cells. Biol Cell 1983; 47:95-110.
279. Raz A, Lotan R. Endogenous galactoside-binding lectins: A new class of functional tumor cell surface molecules related to metastasis. Cancer Metast Rev 1987; 6:433-452.
280. Regan LJ, Dodd J, Barondes SH, Jessel TM. Selective expression of endogenous lactose-binding lectins and lactoseries glycoconjugates in subsets of rat sensory neurons. Proc Natl Acad Sci USA 1986; 83:2248-2252.
281. Jia S, Mee R, Morford G, Argwal N, Voss P, Moutsatsos I, Wang J. Carbohydrate-binding protein 35: molecular cloning and expression of a recombinant polypeptide with lectin activity in Escherichia coli. Gene 1987; 60:197-204.
282. Moutsatsos IK, Wade M, Schindler M, Wang JL. Endogenous lectins from cultured cells: nuclear localization of carbohydrate-binding protein 35 in proliferating 3T3 fibroblasts. Proc Natl Acad Sci USA 1987; 84:6452-6456.
283. Hinek A, Wrenn D, Mecham RP, Barondes SH. The elastin receptor: a galactoside-binding protein. Science 1988; 239:1539-1541.
284. Allen HJ, Karakousis C, Piver MS, Gamarra M, Nava H, Forsyth B, Matecki B, Jazayeri A, Sucato D, Kisailus E, DiCioccio R. Galactoside-binding lectin in human tissues. Tumor Biol 1987; 8:218-229.
285. Carding SR, Thorpe SJ, Thorpe R, Feizi T. Transformation and growth related changes in levels of nuclear and cytoplasmic proteins antigenically related to mammalian β-galactoside-binding lectin. Biochem Biophys Res Commun 1985; 127:680-686.
286. Gabius H-J, Engelhardt R, Cramer F. Endogenous tumor lectins: Overview and perspectives. Anticancer Res 1986; 6:573-578.
287. Gitt MA, Barondes SH. Evidence that a human Soluble β-galacto-

side-binding lectin is encoded by a family of genes. Proc Natl Acad Sci USA 1986; 83:7603-7607.
288. Raz A, Meromsky L, Carmi P, Karakash R, Lotan D, Lotan R. Monoclonal antibodies to endogenous galactoside-specific tumor cell lectins. EMBO J 1984; 3:2979-2983.
289. Raz A, Meromsky L, Lotan R. Differential expression of endogenous lectins on the surface of nontumorigenic, tumorigenic, and metastatic cells. Cancer Res 1986; 46:3667-3672.
290. Raz A, Avivi A, Pazarini G, Garmi P. Cloning and expression of cDNA for two endogenous UV-2237 fibrosarcoma lectin genes. Exp Cell Res 1987; 173:109-116.
291. Raz A, Meromsky L, Zvibel I, Lotan R. Transformation-related changes in the expression of endogenous cell lectins. Int J Cancer 1987; 39:353-360.
292. Raz A, Carmi P, Pazarini G. Expression of two different endogenous galactoside-binding lectins sharing sequence homology. Cancer Res 1988; 48:645-649.
293. Gabius H-J, Engelhardt R, Cramer F, Batge R, Nagel GA. Pattern of endogenous lectins in a human epithelial tumor. Cancer Res 1985; 45:253-257.
294. Irimura T, Matsushita Y, Sutton RC, Carralero D, Ohannesian DW, Cleary KR, Ota DM, Nicolson GL, Lotan R. Increased content of an endogenous lactose-binding lectin in human colorectal carcinoma progressed to metastatic stages. Cancer Res 1991; 51:387-393.
295. Gasic GJ, Gasic TB, Stewart CC. Antimetastatic effects associated with platelet reduction. Proc Natl Acad Sci USA 1968; 61:46-52.
296. Cavanaugh PG, Sloane BF, Honn KV. Role of the coagulation system in tumor cell-induced platelet aggregation and metastasis. Haemostasis 1988; 18:37-46.
297. Grignani G, Jamieson GA. Platelets in tumor metastasis: Generation of adenosine diphosphate by tumor cells is specific but unrelated to metastatic potential. Blood 1988; 71:844-849.
298. Tsuruo T, Watanabe M, Oh-hara T. Stimulation of the growth of metastatic clones of mouse colon adenocarcinoma 26 in vitro by platelet-derived growth factor. Jpn J Cancer Res (Gann) 1989; 80:136-140.
299. Finne J. Enzymic basis for the modulation of cell-surface carbohydrates in lectin-resistant melanoma cells. J Biochem Soc Trans 1983; 11:269-270.
300. Miller FR, Heppner GH. Cellular interactions in metastasis. Cancer Metastasis Rev 1990; 9:21-34.
301. Heppner GH. Tumor cell societies. J Natl Cancer Inst 1989; 81:648-649.
302. Miller BE, Miller FR, Heppner GH. Growth interactions in vivo between tumor subpopulations derived from a single mouse mammary tumor. Cancer Res 1980; 40:3977-3981.
303. Miller BE, Miller FR, Heppner GH. Interactions between tumor cell

subpopulations affecting their sensitivity to the antineoplastic agents cyclophosphamide and methotrexate. Cancer Res 1981; 41:4378-4381.
304. Miller FR. Tumor subpopulation interactions in metastasis. Invasion Metastasis 1983; 3:234-242.
305. Miller FR, Miller BE, Heppner GH. Subpopulation interactions in drug resistance, metastasis and tumor progression. In: Jacquillat C, Weil M, Khayat D, eds. Neo-adjuvant Chemotherapy. London: John Libbey & Co., 1986:69-74.
306. Evans CW. The Metastatic Cell: Behavior and Biochemistry. New York: Chapman and Hall, 1991:551.
307. Steeg PS. Tumor metastasis: before the revolution. Cell 1991; 66:835-836.
308. Fidler IJ, Naito S, Pathak S. Orthotopic implantation is essential for the selection, growth and metastasis of human renal cell cancer in nude mice. Cancer Metastasis Rev 1990; 9:149-165.
309. Hakomori SI. Altered glycosylation in tumor cells. In: Reading ChL, Hakomori SI, Marcus DM, eds. Altered Glycosylation in Tumor Cells. New York: Alan R. Liss, 1988:207-212.
310. Abe K, Hakomori SI, Ohshiba S. Differential expression of difucosyl type 2 chain (LeY) defined by monoclonal antibody AH6 in different locations of colonic epithelia, various histological types of colonic polyps, and adenocarcinomas. Cancer Res 1986; 46:2639-2644.
311. Itzkowitz SH, Yuan M, Gukushi Y, Palekar A, Phelps PC, Shamsuddin AM, Trump BF, Hakomori SI, Kim YS. Lewis X- and sialylated Lewis X-related antigen expression in human malignant and nonmalignant colonic tissues. Cancer Res 1986; 46:2627-2632.
312. Itzkowitz SH, Yuan M, Ferrell LD, Palekar A, Kim YS. Cancer-associated alterations of blood-group antigen expression in human colorectal polyps. Cancer Res 1986; 46:5976-5984.
313. Itzkowitz SH, Yuan M, Montgomery CK, Kjeldsen T, Takahashi HK, Bigbee WL, Kim YS. Expression of Tn, sialosyl Tn, and T antigens in human colon cancer. Cancer Res 1989; 49:197-204.
314. Kim YS, Yuan M, Itzkowitz SH, Sun Q, Kaizu T, Palekar A, Trump BF, Hakomori SI. Expression of LeY and extended LeY blood group-related antigens in human malignant, premalignant, and nonmalignant colonic tissues. Cancer Res 1986; 46:5985-5992.
315. Itzkowitz SH, Bloom EJ, Kokal WA, Modin G, Hakomori S, Kim YS. Sialosyl-Tn: A novel mucin antigen associated with prognosis in colorectal cancer patients. Cancer 1990; 66:1960-1966.
316. Hoff SD, Irimura T, Matsushita Y, Ota DM, Cleary KR, Hakomori S. Metastatic potential of colon carcinoma: Expression of ABO/Lewis-related antigens. Arch Surg 1990; 125:206-209.
317. Gastronovo V, Colin C, Parent B, Foidart JM, Lambotte R, Mahieu, Ph. Possible role of human natural anti-Gal antibodies in the natural antitumor defense system. J Natl Cancer Inst 1989; 81:212-216.
318. Oguchi H, Toyokuni T, Dean B, Ito H, Otsuji E, Jones VL, Sadoizai KK, Hakomori S. Effect of lactose derivatives on metastatic potential of B16 melanoms cells. Cancer Communication 1990; 2:311-316.
319. Glinsky GV. Modification of peptide-protein binding as an extra-

cellular biological function of polyamines: physiological role of primary amines-dependent regulation of the bioactivity of neuropeptides, hormones, growth factors and its significance in the pathogenesis of neoplastic growth. J Tumor Marker Oncology 1987; 1:249-294.
320. Glinsky GV, Linetsky MD, Shemyakin VV, Markin VV, Liventsov VV, Ivanova AB. Structural analysis of a new class of humoral tumor markers: Characterization of 48 aminoglycoconjugates (glycoamines) isolated from rats and human blood plasma. J Tumor Marker Oncology 1990; 5:107-118.
321. Glinsky GV, Linetsky MD. Structural analysis of the carbohydrate core of amino-glycoconjugates (glycoamines) from blood plasma of oncological patients. J Tumor Marker Oncology 1990; 5:137-160.
322. Glinsky GV, Linetsky MD, Liventsov VV, Surmilo NI, Ivanova AB, Sidorenko MV, Vinnitsky VB. Glycoamine-synthetizing reaction: molecular mechanism of the incorporation of monomers into aminoglycoconjugates. J Tumor Marker Oncology 1990; 5:119-136.
323. Linetsky MD, Semenova-Kobzar RA, Glinsky GV. The modifying influence of endogenous biopolymers, glycoamines, from the blood serum of BALB/C mice on aggregation properties of experimental rhabdomyosarcoma cells. Experimental Oncology 1991; 13:6:27-33.
324. Der Hagopian RP, Sugarbaker EV, Ketcham A. Inflammatory oncotaxis. J Am Med Assoc 1978; 240:374-375.
325. Sugarbaker EV, Cohen AM, Ketcham AS. Do metastases metastasize? Ann Surg 1971; 174:161-166.
326. Poste G, Fidler IJ. The pathogenesis of cancer metastasis. Nature 1980; 283:139-146.
327. Weiss L. Metastatic inefficiency. Adv Cancer Res 1990; 54:159-211.
328. Weiss L, Orr FW, Honn KV. Interactions of cancer cells with the microvasculature during metastasis. FASEB J 1988; 2:12-21.
329. Smith RR, Thomas LB, Hilberg AW. Cancer cell contamination of operative wounds. Cancer 1958; 11:53-62.
330. Fisher B, Fisher ER, Feduska N. Trauma and the localization of tumor cells. Cancer 1967; 20:23-30.
331. Glinsky GV. Glycodeterminants of melanoma cell adhesion. A model for antimetastatic drug design. Crit Rev Oncol/Hemat 1993; (in press).
332. Fung PYS, Madej M, Koganty RR, Longenecker BM. Active specific innumotherapy of a murine mammary adenocarcinoma using a synthetic tumor-associated glycoconjugate. Cancer Res 1990; 50:4308-4314.
333. Singhal A, Fohn M, Hakomori SI. Induction of α-N-Acetylgalactosamine-O-Serine/Threonine (Tn) antigen-mediated cellular immune response for active immunotherapy in mice. Cancer Res 1991; 51:1406-1411.
334. Livingston PO, Natoli EJ, Calves MJ, Stockert E, Oettgen HF, Old LJ. Vaccines containing purified GM2 ganglioside elicit GM2 antibodies in melanoma patients. Proc Natl Acad Sci USA 1987;

84:2911-2915.
335. Livingston PO, Ritter G, Srivastava P. Padavan M, Calves MJ, Oettgen HF, Old LJ. Characterization of IgG and IgM antibodies induced in melanoma patients by immunization with purified G_{M2} ganglioside. Cancer Res 1989; 49:7045-7050.
336. Toyokuni T, Dean B, Hakomori SI. Synthetic vaccines: I. Synthesis of multivalent Tn antigen cluster-lysyllysine conjugates. Tetrahedron Lett 1990; 31:2673-2676.
337. Glinsky GV, Linetsky MD, Ivanova AB, Osadchaya LP, Semyonova-Kobzar RA, Surmilo NI. Chemical, biochemical, spectroscopic and biological indentification of structural-functional determinants of glycoamines (aminoglycoconjugates)., J Tumor Marker Oncology 1990; 5:250.
338. Glinsky GV, Ogordniychuk AS, Shilin VV, Linetsky MD, Liventsov VV, Sidorenko MV, Surmilo NI, Kuchar VP. Structural analysis of synthetic structural analogs of glycoamines: contribution into elucidation of the structure and biogenesis of natural aminoglycoconjugates., J Tumor Marker Oncology 1990; 5:250.
339. Ponpipom MM, Bugianesi RL, Robbins JC, Doebber TW, Schen TY. Cell-specific ligands for selective drug delivery to tissues and organs. J Med Chem 1981; 24:1388-95.
340. Watanabe M, Ohishi T, Kuzuoka M, Nudelman ED, Stroud MR, Kubota T, Kodaira S, Abe O, Hirohashi S, Shimosato Y, Hakomori S. In vitro and in vivo antitumor effects of murine monoclonal antibody NCC-SST-421 reacting with dimeric LeA (Leb/LeA) epitope. Cancer Res 1991; 51:2199-2204.
341. Williams WR. The theory of metastasis. In: The Natural History of Cancer. 1908:442-448.
342. Goldmann E. The growth of malignant disease in man and the lower animals, withspecial reference to the vascular system. Proc R Soc Med 1907; 1:1-13.
343. Schmidt MB. In: Fischer J, ed. Die Verbreitungswege der Karzinome und die beziehung generalisiertes sarkome su den leukamischen neubildungen. 1903.
344. Weiss L, Mayhew E, Glaves-Rapp D, Holmes JC. Metastatic inefficiency in mice bearing B16 melanomas. Br J Cancer 1982; 45:44-53.
345. Weiss L. Biomechanical destruction of cancer cells in the heart: a rate regulator for hematogenous metastasis. Invas Metastasis 1988; 8:228-237.
346. Weiss L, Harlos JP, Elkin G. Mechanism of mechanical trauma to Ehrlich ascites tumor cells in vitro and its relationship to rapid intravascular death during metastasis. Int J Cancer 1989; 44:143-148.
347. Weiss L. Metastatic inefficiency: causes and consequences. Cancer Rev 1986; 3:1-24.
348. Weiss L. The hemodynamic destruction of circulating cancer cells. Biorheology 1987; 24:105-115.
349. Weiss L. Biomechanical destruction of cancer cells in the skeletal

muscle: a rate-regulator for hematogenous metastasis. Clin Exp Metast 1989; 7:483-491.

350. Weiss L, Schmid-Schonbein GW. Biomechanical interactions of cancer cells with the microvasculature during metastasis. Cell Biophys 1989; 14:187-215.

351. Orr FW, Adamson IYR, Young L. Pulmonary inflammation generates chemotactic activity for tumor cells and promotes lung metastasis. Am Rev Respir Dis 1985; 131:607-611.

352. Orr FW, Adamson IYR, Young L. Promotion of pulmonary metastasis by bleomycin-induced endothelial injury. Cancer Res 1986; 46:891-897.

353. Glinsky GV. Antigen presentation, aberrant glycosylation and tumor progression. Crit Rev Oncol/Hemat 1994; 17:27-51.

CHAPTER 2

Apoptosis and Metastasis: Failure of Apoptosis in Metastatic Cancer Cells as a Key Factor of Metastatic Disease

2.1. Introduction

Apoptosis (programmed cell death) is an important homeo-static mechanism that balances cell production with cell death and maintains correct cell numbers in the body in physiological and pathological situations.[1-4] Aberrant cell survival resulting from inhibition of apoptosis is expected to contribute to tumor progression and oncogenesis,[3-5] and cancer cells would gain a selective growth advantage by blocking the apoptosis.[3,4,6] The recent findings strongly suggest that apoptosis may also control cell positioning in the body through proper cell-cell and cell-extracellular matrix interactions.[7-10] Matrix-independent survival of metastatic cancer cells during the passage through blood and/or lymph compartments is an essential component of the metastatic cascade that may depend upon high resistance of metastatic cells to apoptosis, since detachment of cells from the extracellular matrix induces programmed cell death[7-10] and malignant transformation confers at least partial resistance to apoptosis induced by disruption of cell-matrix interactions.[7,8] These data support the notion that apoptosis is a mechanism for elimination of misplaced cells and may prevent cancer cells from establishing metastases.[2,4] However, until recently there was no direct experimental evidence linking programmed cell death and metastasis.[20,21,140,143]

Failure of Apoptosis and Cancer Metastasis, by Gennadi V. Glinsky.

Here we will summarize the reports on adhesion-dependent regulation of apoptosis and discuss data indicating that highly metastatic cancer cells exhibit a superior survival ability and resistance to apoptosis induced by diverse external signals.[20,21,140,143] We will also show that higher survival of metastatic cancer cells compared with nonmetastatic or poorly metastatic cells is associated with their enhanced homotypic aggregation and dysfunction of adhesion-dependent regulation of apoptosis. We will overview the recent experimental findings supporting the idea that apoptosis plays a dual role in cancerogenesis, both as a main anticancer defense mechanism and as a major biological selection factor during tumor progression.

Evidence will be considered confirming the notion that apoptosis is an important homeostatic mechanism that controls cell positioning in the body through proper homotypic and heterotypic cell-cell contacts as well as specific cell-extracellular matrix interactions, thereby eliminating misplaced cells and preventing cancer cells from establishing metastases. Along this line, we will identify and discuss numerous factors such as loss or disturbances of cell attachment to ECM, overexpression of certain protooncogenes, auto- and paracrine production of cytokines (α-TNF, β-TGF, etc.), dysfunction of apoptosis regulatory genes, hypoxia, and acidic extracellular pH that may contribute to the activation of apoptosis in malignant tissues in vivo during tumor progression.

Activated apoptosis in cancer tissues will be described as a major biological selection factor during tumor progression that would give no survival alternative during clonal evolution of tumors but the development of apoptosis resistance.[20,21,140,143] The selective pressure of activated apoptosis would lead to the ultimate outcome of malignant progression, namely the emergence of highly metastatic cancer cells resistant to apoptosis. In support of this concept, we will overview several cellular and molecular mechanisms such as overexpression of galactoside-specific lectins and enhanced level of integrins ($\alpha_v\beta_3$; $\alpha_5\beta_1$), dysfunction of apoptosis regulatory genes (*bcl-2* gene family; *p53* and *Rb* genes, focal adhesion kinase (FAK) and nuclear factor kappa B (NF-κB) systems), expression of certain oncogenes (*ras*) that are apparently associated with higher resistance to apoptosis and enhanced metastatic ability of cancer cells. Our analysis will indicate that there is a genetic and molecular basis linking aggressive metastatic phenotype with resistance to apoptosis which would give the metastatic cancer cells an obvious selective growth advantage during tumor progression.

2.2. Adhesion-Dependent Regulation of Apoptosis

The significance of aberrant cellular adhesion for cancer metastasis is widely recognized. Therefore, the analysis of experimental findings on adhesion-dependent regulation of apoptosis may shed some light on the role of programmed cell death in control of metastatic dissemination. A number of recent experimental observations have demonstrated that programmed cell death is an adhesion-dependent process in different cell types, including epithelial cells[8-10] and endothelial cells,[9,11-13] as well as several types of cancer cells.[14-17]

Bates et al[14] reported that survival of a human colon carcinoma cells in vitro is ultimately dependent on their ability to establish and maintain the homotypic intercellular contacts. Human carcinoma cells LIM 1863 grow in vitro as three-dimensional organoids composed of goblet and columnar cells around a central lumen. If organoids are transferred to calcium-depleted medium, they disaggregate into a single cell suspension; reformation of multicellular organoids was observed when the single cells were placed in calcium-containing medium. Reformation of organoids could be blocked by anti-integrin antibodies to the α_v subunit and if single cells were prevented from formation of multicellular organoids, they spontaneously underwent apoptosis.[14] Thus, homotypic cancer cell adhesion may trigger an induction of anti-apoptosis survival mechanism(s) through homotypic intercellular interactions. Heterotypic cell-cell contacts may also function as a cell survival factor and suppressor of apoptosis. The growth and survival of a murine T cell lymphoma cell line CS-21 was reported to depend on intercellular contacts with lymph node stromal cells.[15] Disruption of this type of heterotypic cell-cell adhesion resulted in apoptosis of the lymphoma cells.

Several reports have demonstrated that adhesion of cells to extracellular matrix (ECM) is an important mechanism of inhibition of apoptosis in normal as well as cancer cells.[7-13,16-18] The consensus view is that adhesion to ECM is essential for cell survival via inhibition of programmed cell death, and prevention or disruption of the specific cell-ECM interactions would result in apoptosis. It has been suggested that cell adhesion-dependent regulation of apoptosis is one of the key mechanisms of the anchorage-dependency of growth which is disrupted in cancer cells.[7] ECM-dependent inhibition of apoptosis is likely to be mediated by integrin-activated signaling pathway modulating the activity of apoptosis regulatory genes such as *bcl-2* or interleukin-1 β converting enzyme(ICE)-like proteases. Upregulation of apoptosis-suppressor *bcl-2*[18] and inhibition of apoptosis-activator ICE-like protease(s)[10] was shown to be involved in integrin-mediated suppression of apoptosis in certain cell types. Malignant transformation confers at least partial resistance to apoptosis induced by aberrations in cellular adhesion, since expression of certain oncogenes was shown to cause a subversion of physiological mechanisms of adhesion-dependent regulation of apoptosis.[7,8,19]

Taken together these data strongly support the notion that apoptosis is an important homeostatic mechanism that controls cell positioning in the body through proper homotypic and heterotypic cell-cell contacts as well as specific cell-ECM interactions.[7-10,20-23] These findings are also in agreement with the idea that apoptosis is a mechanism for elimination of misplaced cells that should prevent cancer cells from establishing metastases.[2,4] However, what is known regarding susceptibility of metastatic cancer cells to apoptosis?

2.3. Superior Resistance of Metastatic Cancer Cells to Programmed Cell Death

We compared the response of several poorly and highly metastatic cell lines to different external signals leading to apoptosis.[20,21,140,143] These cell lines were selected for high metastatic potential through completely unrelated in vivo and in vitro procedures.[24-26] These cell lines possess very similar growth characteristics; however, they have a clearly distinct malignant potential in vivo and they differentially responded to high-saturation density growth or serum deprivation, conditions known to induce apoptosis in many other cell lines,[27] including different transformed cell lines.[28,29] Induction of cell death in the different metastatic cell lines upon growth from low to high density or serum deprivation was monitored in a viability assay based on trypan blue and/or acridine orange staining. Highly metastatic cell lines maintained viability for several days after reaching saturation density or following serum deprivation. In contrast, poorly metastatic cells demonstrated a pronounced and rapid loss of viability after achieving saturation density or following transfer into medium containing a suboptimal concentration (1% or 0.1%) of fetal calf serum.

Cell death can occur nonspecifically by necrosis or by a specific process called apoptosis.[1-5,30] Main features that distinguish programmed cell death from necrosis are the induction of nuclear DNA fragmentation in intranucleosomal regions (characteristic ladder pattern of DNA fragmentation) and chromatin condensation.[30] By using these criteria, dying metastatic cell lines were examined for induction of apoptosis before and during maximal differential viability of the cell lines. High levels of DNA fragmentation with characteristic ladder patterns, diagnostic of apoptosis and indicative of intranucleosomal cleavage, occurred in dying cells from poorly metastatic cell lines.[20,21,140] Significantly smaller, or no, amounts of low molecular weight DNA were detectable in identically treated highly metastatic cell lines. Identification of apoptotic cells by the terminal deoxynucleotidyl transferase-mediated dUTP-X nick end labeling method (the TUNEL method) and acridine orange staining was consistent with DNA fragmentation analysis.[20,21]

Thus, highly metastatic cells (B16 F10 and B16 BL6 murine melanoma) exhibited a superior survival ability and resistance to apoptosis compared to poorly metastatic B16 F1 cells. This phenomenon is not restricted to the murine B16 melanoma experimental metastasis model. In a similar experimental setting, two human cancer cell lines highly metastatic in nude mice (breast carcinoma cell line MDA-MB-435 and melanoma cell line TXM-13) showed superior viability and resistance to apoptosis compared to their poorly metastatic counterparts such as breast carcinoma cell lines MDA-MB-468, MDA-MB-361, and MDA-MB-231, as well as melanoma cell line DX-3.[20,21,140]

Differential sensitivity to apoptosis of nonmetastatic and metastatic cancer cells is not limited only to saturation density- and/or serum deprivation-induced programmed cell death. It has been reported that metastatic and nonmetastatic K1735 melanoma cell variants differ in their susceptibility to Fas-mediated apoptosis.[31] The H7 hybrid cells, formed by fusing murine nonmetastatic K1735 C19H melanoma cells with human highly metastatic A375 C15 melanoma cells exhibited high metastatic potential, retained human chromosome 17, and expressed mutated p53 protein.[32,33] Both H7 cells as well as K1735 C19H cells express similar levels of Fas/APO-1 receptor which triggers apoptosis upon binding of Fas ligand or anti-Fas antibody. However, only 7% of H7 cells were killed in the presence of anti-Fas antibody, compared with 50% killing of K1735 C19H cells.[31] Since Fas ligand is known to be constitutively expressed in the lungs, small intestine, ovary and testes, the Fas:Fas ligand interaction-mediated induction of apoptosis could be regarded as an important component of organ-specific anti-metastatic mechanism(s).

It was also reported that highly metastatic human breast carcinoma cells MDA-MB-435 is significantly more resistant to apoptosis induced by doxorubicin, staurosporine or taxol C when compared with poorly metastatic MDA-MB-468 cells.[34] The level of apoptosis as assessed by DNA laddering and fragmentation assays, flow cytometry gated for sub-G_1 nuclei, and morphological changes ascertained by transmission electron microscopy (blebbing, chromatin condensation, cellular shrinkage) was significantly diminished in drug-treated MDA-MB-435 cells compared with the identically treated MDA-MB-468 cells. Importantly, the apoptosis resistance of MDA-MB-435 cells was associated with higher chemoresistance to staurosporine and doxorubicin.[34] Poorly metastatic LNCaP-Pro5 human prostate carcinoma cells were significantly more sensitive to thapsigargin-induced apoptosis than were highly metastatic LNCaP-LN3 cells,[141] as determined by viability assay, DNA fragmentation analysis, and caspase-mediated cleavage of the DNA-repair enzyme, poly(ADP-ribose)polymerase.

Thus, highly metastatic cancer cells exhibited significantly higher resistance to apoptosis induced by diverse external signals when compared with nonmetastatic or poorly metastatic cancer cells. This superior resistance to apoptosis would give the metastatic cancer cells an obvious selective growth advantage during tumor progression. What are the possible cellular and molecular mechanisms of the higher resistance of metastatic cancer cells to programmed cell death?

2.4. Cellular and Molecular Mechanisms of Higher Resistance to Apoptosis by Metastatic Cancer Cells

2.4.1. Galactoside-Specific Lectins as Apoptosis Inhibitors in Metastatic Cancer Cells

In several experimental systems, including the murine B16 melanoma experimental metastasis model, increased levels of β-galactoside-binding lectins have been associated with an enhanced ability of cancer cells for homotypic and heterotypic aggregation, anchorage independent growth, and metastatic propensity.[35-38] Inhibition of these activities and of metastasis by antibodies against lectins and galactose-rich glycoproteins further supports the involvement of β-galactoside-specific lectins in cancer cell aggregation, adhesion, and metastatic spreading.[39,40] We reasoned that matrix-independent survival of metastatic cells might be associated with their homotypic and heterotypic aggregation ability,[35] mediated by specific carbohydrate-lectin recognition and binding. Indeed, highly metastatic cancer cells showed significantly increased levels of both spontaneous and asialofetuin-induced homotypic aggregation,[20,21] which has been shown to be mediated through carbohydrate-lectin recognition and binding.[36-40] Addition of the cell aggregation promoter asialofetuin to culture medium enhanced the β-galactoside-specific lectin-dependent homotypic aggregation of poorly metastatic B16-F1 melanoma cells, significantly increased their survival under the conditions of serum deprivation or saturation density, and inhibited induction of apoptosis.[20,21] Moreover, disruption of β-galectin-mediated homotypic aggregation of highly metastatic MDA-MB-435 human breast carcinoma cells, as well as murine B16-F10 melanoma cells, by anti-metastatic synthetic glycoamines caused an induction of apoptosis.[20-23]

Therefore, carbohydrate-lectin recognition and binding, displayed phenotypically in homotypic cancer cell aggregation, appears to be an important factor linked with superior resistance of metastatic cells to apoptosis. It seems likely that yet unidentified lectin-associated signaling pathway may lead to activation of the anti-apoptosis gene(s) and/or inhibition of the apoptosis activator gene(s) in metastatic cancer cells. Consistent with this notion, the expression of endogenous tumor-associated β-galactoside-specific lectin (galectin-3), which acts as a receptor for ligands containing poly-N-acetyllactosamine structures, has been correlated with neoplastic progression in the colon[41] and with the malignant potential of primary human brain tumors.[42] Furthermore, expression of galectin-3 correlated with the metastatic potential of the colon cancer cells, and cancer metastases to the CNS were all moderately to strongly positive for galectin-3 expression.[42,43] It was also reported that tissue levels of 31 kDa lactose-binding lectin were higher in primary gastric carcinomas and their metastases compared with adjacent normal gastric mucosa.[116]

It is well established that carbohydrate-based adhesion of metastatic cancer cells is an important mechanism of their homotypic and heterotypic aggregation in blood and/or lymph of tumor hosts[35-40] and may play a key role in mediating the binding of metastatic cancer cells to blood vessel endothelia.[35,44-46] Recent data indicated that carbohydrate-mediated adhesion could be a major survival factor contributing to superior resistance to apoptosis by metastatic cancer cells.[20,21,143] Therefore, the inhibitors of carbohydrate-based adhesion that have been suggested as prospective therapeutic agents for a wide range of human diseases[35,45] also may act as inducers of apoptosis in target cells, which would significantly extend the area of their potential therapeutic applications.

Recently we have extended our initial findings on high resistance to apoptosis of metastatic cancer cells. We found that highly metastatic human melanoma and breast carcinoma cells also manifest an increased resistance to apoptosis.[140] High levels of β-galectin expression and increased homotypic aggregation were associated with the increased apoptosis resistance of highly metastatic MDA-MB-435 human breast carcinoma cells as well as human prostate carcinoma cell line PC3M-LN4. The ability of galectin-3 to inhibit apoptosis has been demonstrated directly in gene-transfer experiments utilizing human leukemia cell line Jurkat E6-1.[142] It has been suggested that galectin-3 may function as an apoptosis regulator through a cell death inhibition pathway involving Bcl-2,[142] since galectin-3 can directly interact with Bcl-2 and form galectin-3/Bcl-2 heterodimers.

2.4.2. Cell Receptor Molecules, Intracellular Signaling and Deficiency of Apoptosis Execution

It should be noted that recent reports demonstrate the central role of focal adhesion kinase (FAK), an intracellular signaling molecule associated with the intracellular tails of integrins and interacting with various cytoskeletal proteins, in determining the resistance to apoptosis of detached cells. Canine and human epithelial cells can survive without contact with the extracellular matrix if they express activated FAK, in contrast to the same cell lines that were not engineered to constitutively express FAK and underwent apoptosis.[125] These data suggest that activation of FAK helps cells sense attachment to the extracellular matrix and initiates the antiapoptotic intracellular signaling mechanism essential for cell survival. This mechanism may well be relevant to the increased resistance of highly metastatic cancer cells to apoptosis, since it was demonstrated that FAK is overexpressed in metastatic human breast and colon cancers, that reduction of the expression of FAK using antisense oligonucleotides causes apoptosis in cultured cells derived from human rhabdomyosarcoma, and that FAK overexpression correlates with the ability of cancer cells to metastasize.[126]

Our preliminary data indicate that highly metastatic cancer cells are resistant to the tumor necrosis factor-α (TNF-α)-induced apoptosis. It has been discovered recently that resistance to the TNF-α-induced apoptosis of several cell types, including cancer cells, is mediated by a protein called nuclear factor kappa B (NF-κB) and its inhibitor IκBα.[127-129] It would be of interest to determine if the NF-κB system is responsible for high resistance of metastatic cancer cells to TNF-α-induced apoptosis, and other forms of apoptosis resistance as well. Many signal transduction pathways leading to NF-κB activation result in an inducible phosphorylation, ubiquitination and subsequent degradation by the proteasome of the NF-κB inhibitor, IκBα.[130-133] Thus, one possible molecular option would be that activation of FAK may induce serine phosphorylation of IκBα on residues 32 and 36 (or on the COOH-terminal PEST region), leading to degradation of IκBα, activation of NF-κB and inhibition of apoptosis.

Our experimental observations clearly demonstrate that several key apoptosis effector mechanisms are deficient in highly metastatic cancer cells. For example, highly metastatic cancer cells exhibit significantly lower levels of nuclear Ca^{2+}-dependent endonuclease activity as well as diminished basal and inducible activities of ICE-like proteases (caspases) compared to their poorly metastatic counterparts,[140] which evidently may contribute to the increased apoptosis resistance. The intriguing problem is whether this profound deficiency in apoptosis execution mechanisms is primarily responsible for apoptosis resistance or it just reflects some genetic alterations at the higher levels of apoptosis control such as FAK activity, Bcl-2 family balance, NF-κB system operation, p53 and Rb status, etc.

As we already mentioned, the expression and function of specific integrins such as $\alpha_v\beta_3$ and $\alpha_5\beta_1$ were correlated with resistance to apoptosis by certain cell types, and disruption of integrin-mediated adhesion induces apoptosis.[7-13,16-18] For example, the integrin $\alpha_v\beta_3$ rescues melanoma cells from apoptosis in three-dimensional dermal collagen cultures.[17] It is important to note, that critical involvement of integrin $\alpha_V\beta_3$ in melanoma tumorigenicity, progression, and metastasis has been reported earlier.[117-121] Similarly, spreading and migration of carcinoma cells on vitronectin and fibrinogen have been shown to require the expression of $\alpha_V\beta_3$ integrin.[121,122] Finally, it has been reported recently that a high level of expression of surface integrin $\alpha_V\beta_3$ has an important role in cell adhesion, motility, and liver metastasis of murine RAW117 large cell lymphoma.[123]

2.4.3. The Balance of Apoptosis Regulatory Proteins in Metastatic Cancer Cells

There is some initial experimental evidence indicating that higher resistance to apoptosis of metastatic cancer cells is associated with mutations and/or dysfunction of the apoptosis regulatory genes. It was shown that increased expression of Bcl-2, and especially Bcl-X_L, is associated with apoptosis resistance and chemoresistance of highly metastatic MDA-MB-435 human breast carcinoma cells,[34] whereas the levels of expression of Bax appear similar between the highly and poorly metastatic human breast carcinoma cells. Our preliminary immunoblot experiments show that highly metastatic human breast carcinoma cells MDA-MB-435 express higher levels of apoptosis inhibitors Bcl-2 and Bcl-X_L, as well as lower levels of proapoptotic protein Bax compared to poorly metastatic MDA-MB-468 cell line. Apoptosis resistance of highly metastatic human prostate carcinoma cell line LNCaP-LN3 was associated with high level of expression of the cell death inhibitor Bcl-2 and low level of the cell death promoters Bax and Bak.[141] It would be of interest to investigate whether this type of disbalance in the Bcl-2 family of apoptosis regulatory proteins is characteristic of other highly metastatic cancer cells as well, and whether it could account for their increased resistance to apoptosis. Correspondingly, it was previously reported that experimental overexpression of the Bcl-2 or the Bcl-X_L proteins in cancer cells can increase their resistance to a variety of anti-cancer drugs.[47-49] It has also been shown that overexpression of Bcl-2 protein provides protection against a wide variety of apoptotic signals including growth factor deprivation, loss of cell attachment to the extracellular matrix, oncogenes (c-myc), tumor suppressor genes (p53), cytolytic T cells, radiation, and chemotherapeutic agents.[6,7,50]

2.4.4. *Sensitivity to Apoptosis and Multicellular Drug Resistance*

Resistance to chemotherapy remains one of the most significant problems in treating cancer patients. Recently a new concept of acquired drug resistance in tumors was developed based on the response of a cell population (multicellular chemoresistance) as opposed to classic unicellular resistance mechanisms.[134,135] Application of inhibitors of cancer cell adhesion was suggested as an adjuvant treatment strategy to reverse this type of chemoresistance in tumors.[136-138] In our collaborative study with A. Frenkel and R.S. Kerbel, we have investigated the ability of antimetastatic synthetic glycoamines to cause a reversal of multicellular drug resistance in tumors.[139] Upon exposure to taxol for 48 hours, a panel of human ovarian cancer cell lines demonstrated a drug-resistant phenotype when grown as three-dimensional cultures (multicellular tumor spheroids), but did not exhibit this phenotype when grown in monolayer cultures. After treatment with taxol, the cells grown as three-dimensional cultures had a greater survival and increased proliferative potential as well as colony forming ability compared to the taxol-exposed cells grown in monolayer cultures. Taxol-induced apoptosis in the monolayer cultures as measured by flow cytometry for concomitant detection of apoptosis and cell cycle analysis was not observed in the spheroids. We have studied the ability of synthetic glycoamines, antiadhesive and animetastatic agents, to reverse the multicellular resistance of ovarian cancer cells to taxol-induced apoptosis as well as to synergize with cisplatin in inducing apoptosis in ovarian cancer cells. These antimetastatic glycoamine analogs act as inhibitors of multicellular tumor spheroid formation by competing for specific carbohydrate-lectin interactions, particularly those involving β-galactoside-specific lectins expressed on tumor cells.[22,23] Three synthetic glycoamines (Fru-Gly, Fru-D-Leu, and Lac-L-Leu) induced programmed cell death in the ovarian cancer cell line A2780. These compounds clearly synergized in vitro with taxol and cisplatin to increase the apoptotic index of ovarian cancer cells, as well as causing a reversal of multicellular resistance to taxol of ovarian cancer cells. Our preliminary observations indicate the potential role of antiapoptotic protein Bcl-X_L in the multicellular drug resistance phenotype.[139] The level of antiapoptotic protein Bcl-X_L decreased significantly upon taxol treatment in the monolayer cultures, whereas the Bcl-X_L levels in the spheroids treated with taxol remained elevated, closely resembling the levels of the control spheroids. However, the treatment of ovarian cancer cells grown as multicellular tumor spheroids with synthetic glycoamines resulted in significant reduction of Bcl-X_L levels, providing a reasonable molecular model for mechanisms of synergy of antimetastatic synthetic glycoamines with chemotherapeutic drugs in inducing apoptosis in cancer cells. Our results support the hypothesis that antiadhesive and antimetastatic agents—synthetic glycoamines—may have a translational potential in combination with chemotherapeutic drugs to reverse multicellular drug resistance and to improve the efficacy of treatment of cancer patients.

2.4.5. *Anti-Oncogene and Oncogene Expression*

Recent analysis of the aforementioned H7 cells revealed that mutations at codons 249 and 273 of the human p53 gene confer metastatic potential to nonmetastatic mouse melanoma cells.[51] Thus, mutations in the p53 gene may be responsible for higher resistance to the Fas-induced apoptosis of metastatic H7 cells compared with the parental nonmetastatic K1735 C18H cell line. It was reported earlier that the presence of a mutant p53 gene or the absence of wild-type p53 alleles is associated with increased resistance of cancer cells to apoptosis induced by DNA-damaging agents such as chemotherapy or radiation.[52-54] Thus, both activation of apoptosis-suppressor genes as well as inhibition of apoptosis-activator genes may confer resistance to the programmed cell death of metastatic cancer cells.

Expression of certain oncogenes, particularly those that have been linked previously with the enhanced metastatic potential of cancer cells and poor prognosis, may also contribute to the aberrant cell survival and higher resistance to apoptosis. There are several well documented examples of promoting cell survival by the *ras* family of oncogenes: (1) there is an association between *ras* oncogene expression and reduced levels of apoptosis in vivo in transplantable mouse solid tumors;[55-57] (2) apoptosis induced by disruption of epithelial cell-ECM interactions can be inhibited by expression of the v-H-*ras* oncogene;[8] (3) mutant H-*ras* can prevent E1A-induced apoptosis in primary BRK cells;[58] (4) apoptosis in intestinal epithelial cells induced by three-dimensional growth conditions could be suppressed by mutant c-H-*ras* oncogene expression.[19] It has been suggested that mutant *ras* oncogenes may contribute to carcinogenesis and tumor progression, particularly during development of colon cancer, not only as a promoter of uncontrolled cell proliferation, but also as an inhibitor of apoptosis, thus facilitating an enhanced cancer cell survival.[19] At least in some cases, the anti-apoptotic effects of mutant ras oncogenes may be mediated through members of the *bcl-2* gene family.[59]

2.5. Role of Apoptosis in Tumor Progression

It has been suggested that apoptosis is a mechanism for eliminating cells that have been produced in excess, that have developed improperly, or that have sustained genetic damage.[57,60] One might expect, therefore, that apoptosis should be activated in malignant tumors. The concept that apoptosis plays a significant homeostatic role in neoplasia is supported by the observations that identified the increased apoptosis in a range of experimental tumors.[61] The progressive rise of apoptosis level from premalignant to malignant tumors as well as with neoplastic lesion grade increase, was detected in several types of human cancer, including liver,[62] prostate,[63,64] cervical neoplasias,[65] and squamous carcinomas.[66] Thus, cancer cells are susceptible to apoptosis in vivo and they readily undergo apoptosis during tumor development, conceivably as a consequence of a constant selective pressure of various extrinsic and intrinsic pro-apoptotic signals. It follows that activation of a range of pro-apoptosis signals and induction of apoptosis should be regarded as a major factor of biological selection during tumor progression. Considering the apparent genetic link between resistance to apoptosis and metastatic phenotype (see above), the ultimate result of tumor progression would be the selection of metastatic cancer cells resistant to apoptosis. Indeed, there is some initial experimental evidence confirming the selection of cancer cells resistant to apoptosis during development of colonic neoplasia as well as breast carcinoma.

In vivo expansion of neoplastic colonic epithelial cells results both from increased rates of cell proliferation and from decreased rates of cell death due to apoptosis.[61,67,68] Decreased proportions of apoptotic cells have been detected during progression of colonic neoplasia from benign adenomatous polyps to invasive cancers, and there seems to be a direct correlation between the susceptibility of epithelial cells to undergo apoptosis and the stage of disease progression.[69] Shibata et al[124] studied the alterations in apoptosis during mammary tumor progression in C3(1)/SV40 large T antigen transgenic mice. Apoptosis levels, assessed by in situ end labeling, were low in normal mammary epithelial cells, significantly higher in adenocarcinomas, and at highest levels in preneoplastic lesions (atypical hyperplasias). The workers concluded that marked activation of apoptosis in preneoplastic lesions and relative suppression of apoptosis during the transition from atypical hyperplasia to adenocarcinoma is a critical event for mammary cancer development in C3(1)/T_{AG} transgenic mice.[124]

Recent analysis revealed some molecular mechanisms of the acquired resistance to apoptosis during progression of colorectal cancers. The expression of several Bcl-2 family members (including Bcl-2, Bcl-X, Mcl-1, Bax, and Bak) was characterized in a group of colorectal adenocarcinomas and adenomatous polyps.[70] The intensity of Bcl-X immunostaining was elevated in 60% of carcinomas and 50% of adenomatous polyps, and immunoblot analysis indicated a marked elevation in the relative levels of the anti-apoptotic Bcl-X_L protein in tumors compared to adjacent normal colonic tissues. Furthermore, the intensity of immunostaining for the pro-apoptotic protein Bak was reduced compared to that of normal mucosal epithelial cells in 90% of carcinomas and 92% of adenomatous polyps, suggesting that reduction in Bak expression may occur early in colorectal tumor progression.[70] Thus, decreased expression of the apoptosis promoter Bak and enhanced expression of the apoptosis blocker Bcl-X_L may contribute to the decreased rates of apoptosis during progression of colorectal neoplasia. The decreases in Bcl-2 and Mcl-1 expression were characterized as a later event associated with progression of colorectal cancers, whereas Bax expression was relatively constant. It has been suggested, therefore, that a shift in expression of anti-apoptotic proteins (from early Bcl-2 and Mcl-1 expression to Bcl-X_L overexpression), as well as decreased expression of the pro-apoptotic protein Bak, may occur during colorectal cancer progression.[70]

The Bcl-2 family of proteins is likely to play a key role in higher resistance of metastatic cancer cells to apoptosis. The members of this gene family regulate a critical distal step of the programmed cell death pathway and are represented by both anti-apoptotic proteins such as Bcl-2, Bcl-X_L, and Mcl-1 as well as pro-apoptotic effectors such as Bax, Bak, and Bcl-X_S.[2,6,71-82] These proteins can directly interact with each other in a network of homo- and heterodimers, and the relative balance of apoptosis inhibitors and apoptosis activators would determine the sensitivity of a given cell to apoptosis. Bcl-2 and Bcl-X_L seem to inhibit apoptosis through a common pathway[83] and they both are functional targets of the pro-apoptosis protein, Bcl-X_S.

Bcl-2 has been shown to be overexpressed in a range of human tumors, including up to 70% of breast cancers,[84] and Bcl-X_L is overexpressed in a significant percentage of primary cell lines and tissues derived from breast cancer.[85] However, at least in some types of solid tumors such as breast cancer and melanoma, high levels of Bcl-2 are associated with a more favorable prognosis.[86] Thus, some other Bcl-2 family member may play a role in development of apoptosis resistance in these types of tumors. Indeed, frequent reduction in Bax immunostaining was reported recently in breast adenocarcinomas,[87,88] and reduced expression of the pro-apoptotic gene *bax* was associated with poor response rates to combination chemotherapy and shorter survival in women with metastatic breast adenocarcinoma.[87]

It is well established that overexpression of certain protooncogenes such as *c-myc*, *c-myb*, and the *E2F* family members induces apoptosis in the presence of functional p53. Although the mechanism by which oncogenes cooperate with p53 in promoting apoptosis remains obscure, recent data indicate direct involvement of the Bcl-2 family member Bax. It has recently been shown that p53 regulates *bax* gene expression by directly interacting and transactivating its 5'-flanking region.[89] Overexpression of the *c-myb* protooncogene strongly stimulates p53-mediated activation of the *bax* promoter and further increases the level of *bax* mRNA expression, and this effect was correlated with enhanced apoptosis of human and murine cell lines.[90] Overexpression of another tumor-suppressor gene product, retinoblastoma (RB) protein, has recently been shown to prevent cells from undergoing p53-dependent apoptosis.[91,92] It has been suggested that the hypophosphorylated form of RB is a target for cleavage by ICE-like proteases during the transition from commitment to the execution stage of apoptosis and the cleavage of RB might be involved in determination of the decision fork between proliferation and apoptosis.[93] It has also been shown that apoptosis of prostate epithelium is mediated by a regulatory mechanism requiring active RB, and in the human prostate cancer cell line LNCaP apoptosis was preceded by dephosphorylation of RB protein.[94]

Acidic extracellular pH may also contribute to the activation of apoptosis in malignant tumors in vivo. As a consequence of the elevated aerobic glycolysis and reduced blood flow[95-99] the extracellular pH of tumors is generally more acidic than that of normal tissues.[100-102] It was observed that acidic pH increases metastatic potential of implanted murine sarcoma, melanoma, and lymphoma cells[103] and enhances the invasive behavior of human melanoma cells.[104] Furthermore, murine B16-F10 melanoma cells incubated in an acidic medium demonstrated increased migratory ability through type IV collagen gel compared with cells grown in a neutral medium.[105] A relatively mild acidic environment similar to that observed in tumors in vivo causes apoptosis in a range of cancer cell lines.[106] Apoptosis began to appear as judged with the TUNEL staining and electron microscopy when medium pH was lowered to 6.8-7.0, peaking at medium pH 6.2-6.4, at which point intracellular pH was 6.5-7.0. Hypoxia was shown to induce apoptosis in cancer cells; this was predominantly prevented by Bcl-2 and Bcl-X_L expression.[144] This mechanism may be relevant to the hypoxia-mediated selection of cancer cells with diminished apoptotic potential in solid tumors.[145] Hypoxia is considered one of the key microenvironmental factors in the development of solid tumors in vivo.[146]

Thus, numerous factors such as loss or disturbances of cell attachment to extracellular matrix, overexpression of certain protooncogenes, auto- and paracrine production of cytokines (α-TNF, β-TGF, etc.), dysfunction of apoptosis regulatory genes, hypoxia, and acidic extracellular pH may apparently contribute to the activation of apoptosis in malignant tumors in vivo. Activated apoptosis in cancer tissues should be regarded as a major biological selection factor during tumor progression that would give no survival alternative during clonal evolution of tumors except development of apoptosis resistance, and would lead to the ultimate outcome of malignant progression, namely the emergence of resistant to apoptosis metastatic cancer cells.

2.6. Conclusion

The genetic and molecular basis for the association of metastatic phenotype with resistance to apoptosis is the intriguing subject of further investigations. However, selection of cancer cells with enhanced metastatic potential results in coselection of highly metastatic cells with increased resistance to programmed cell death. Is the opposite also correct, that selection of apoptosis-resistant cancer cells may be associated with the enhancement of their metastatic propensity? If so, the detection of apoptosis-resistant cancer cells and the analysis of the expression of anti-apoptosis gene(s) such as *bcl-2* would have obvious prognostic and therapeutic significance. Indeed, the *bcl-2* expression has been associated with a poor prognosis in neuroblastoma,[107] prostate cancer,[108] and colon cancer.[109] However, at least in some types of solid tumors such as breast cancer and melanoma, high levels of Bcl-2 were reported to be associated with a more favorable prognosis.[86] Thus, some other Bcl-2 family member may play a role in development of apoptosis resistance in these types of tumors. Indeed, frequent reduction in Bax immunostaining was reported in breast adenocarcinomas,[87,88] and reduced expression of the pro-apoptotic gene *bax* was associated with poor response to combination chemotherapy and shorter survival in women with metastatic breast adenocarcinoma.[87] As we already have indicated, the members of this gene family regulate a critical distal step of the programmed cell death pathway and are represented by both anti-apoptotic proteins such as Bcl-2, Bcl-X_L, and Mcl-1 as well as pro-apoptotic effectors such as Bax, Bak, and Bcl-X_S.[2,6,71-82] These apoptosis regulatory proteins can directly interact with each other in a network of homo- and heterodimers, and the balance of apoptosis inhibitors and apoptosis activators would determine the sensitivity or resistance of a given cell to apoptosis. Therefore, the simultaneous determination of the levels of several proteins of the Bcl-2 family network seems to be the most productive experimental approach.[70] Recent experimental observations indicated that certain signal transduction mechanisms such as serine/threonine phosphorylation and cleavage by apoptosis-associated proteases can significantly affect apoptosis-modification activities of Bcl-2 family members.[147]

Moreover, the identification and cellular monitoring of an apoptosis-related gene(s) may help to develop an efficient genetic marker(s) of metastatic cells, and searches for chemical inducers of apoptosis in cancer cells could lead to discoveries of long-awaited effective anti-metastatic drugs. The data summarized here have also important fundamental implications. The association of the aggressive metastatic phenotype with resistance to apoptosis provides a plausible explanation to the long-standing fundamental problem which is critical for our understanding of cancer: why would clonal evolution and clonal selection during tumor progression inevitably result in metastasis? Finally, the cell death that is induced by gamma- or X-irradiation, as well as by a wide range of cytotoxic drugs used in cancer chemotherapy, has been found to be apoptosis.[110-114] Therefore, the apparent link between high metastatic potential and resistance to programmed cell death may explain the selection of highly aggressive metastatic cancer cells during unsuccessful cases of clinical cancer therapy. In agreement with this notion, the overexpression of the apoptosis-suppressor gene Bcl-X_L was detected in highly metastatic cancer cells[34] and Bcl-X_L overexpression was shown to confer a multidrug resistance phenotype.[115] Resistance to chemotherapy remains one of the most significant problems in treating cancer patients. Further elucidation of the apparent molecular overlap between resistance to apoptosis of metastatic cancer cells and their chemoresistance may help to develop an effective novel therapeutic strategy to reverse the multicellular drug resistance and to improve the efficacy of treatment of cancer patients.

References

1. Wyllie AH, Kerr JFR, Currie AR. Cell death: The significance of apoptosis. Int Rev Cytol 1980; 68:251-306.
2. Thompson CB. Apoptosis in the pathogenesis and treatment of disease. Science 1995; 267:1456-1462.
3. Williams GT, Smith CA. Molecular regulation on apoptosis: Genetic control on cell death. Cell 1993; 74:777-779.
4. Raff MC. Social control on cell survival and cell death. Nature 1992; 356:397-400.
5. Williams GT. Programmed cell death: Apoptosis and oncogenesis. Cell 1991; 65:1097-1098.
6. Reed JC. Bcl-2 and the regulation of programmed cell death. J Cell Biol 1994; 124:1-6.
7. Ruoslahti E, Reed JC. Anchorage dependence, integrins, and apoptosis. Cell 1994; 77:477-478.
8. Frisch SM, Francis H. Disruption of epithelial cell-matrix interactions induces apoptosis. J Cell Biol 1994; 124:619-626.
9. Meredith JE, Fazeli B, Schwartz MA. The extracellular matrix as a cell survival factor. Mol Biol Cell 1993; 4:953-961.
10. Boudreau N, Sympson CJ, Werb Z, Bissell MJ. Suppression of ICE and apoptosis in mammary epithelial cells by extracellular matrix. Science 1995; 267:891-893.
11. Brooks PC, Clark RAF, Cheresh DA. Requirement of vascular integrin $\alpha_v\beta_3$ for angiogenesis. Science 1994; 264:569-572.
12. Brooks PC, Montgomery AMP, Rosenfeld M, Reisfeld RA, Hu T, Klier G, Cheresh DA. Integrin $\alpha_v\beta_3$ antagonists promote tumor regres-

sion by inducing apoptosis of angiogenic blood vessels. Cell 1994; 79:1157-1164.
13. Re F, Zanetti A, Sironi M, Polentarutti N, Lanfrancone L, Dejana E, Colotta F. Inhibition of anchorage-dependent cell spreading triggers apoptosis in cultured human endothelial cells. J Cell Biol 1994; 127:537-546.
14. Bates RC, Buret A, van Helden DF, Horton MA, Burns GF. Apoptosis induced by inhibition of intercellular contact. J Cell Biol 1994; 125:403-415.
15. Fujita N, Kataoka S, Naito M, Heike Y, Boku N, Nakajima M, Tsuruo T. Suppression of T-lymphoma cell apoptosis by monoclonal antibodies raised against cell surface adhesion molecules. Cancer Res 1993; 53:5022-5027.
16. Kim WH, Schnaper HW, Nomizu M, Yamada Y, Kleinman HK. Apoptosis in human fibrosarcoma cells is induced by a multimeric synthetic tyr-ile-gly-ser-arg (YIGSR)-containing polypeptide from laminin. Cancer Res 1994; 54:5005-5010.
17. Montgomery AMP, Reisfeld RA, Cheresh DA. Integrin $\alpha_v\beta 3$ rescues melanoma cells from apoptosis in three-dimensional dermal collagen. Proc Natl Acad Sci USA 1994; 91:8856-8860.
18. Zhang Z, Vuori K, Reed JC, Ruoslahti E. The $\alpha 5\beta 1$ integrin supports survival of cells on fibronectin and up-regulates Bcl-2 expession. Proc Natl Acad Sci USA 1995; 92:6161-6165.
19. Rak J, Mitsuhashi Y, Erdos V, Huang S, Filmus J, Kerbel RS. Massive programmed cell death in intestinal epithelial cells induced by three-dimensional growth conditions: suppression by mutant c-H-ras oncogene expression. J Cell Biol 1995; 131:1587-1598.
20. Glinsky GV, Glinsky VV. Apoptosis and metastasis: a superior resistance of metastatic cancer cells to programmed cell death. Cancer Letters 1996; 101:43-51.
21. Glinsky GV, Glinsky VV, Price JE. Association of aggressive metastatic phenotype with higher resistance to apoptosis. Proc Amer Assoc Cancer Res 1996; 37:14.
22. Glinsky GV, Mossine VV, Price JE, Bielenberg D, Glinsky VV, Ananthaswamy HN, Feather MS. Inhibition of colony formation in agarose of metastatic human breast carcinoma and melanoma cells by synthetic glycoamine analogs. Clin Exp Metastasis 1996; 14:253-267.
23. Glinsky GV, Price JE, Glinsky VV, Kiriakova G, Mossine VV, Metcalf JB. Inhibition of human breast cancer metastasis in nude mice by synthetic glycoamines. Cancer Res 1996; 56:5325-5329.
24. Fidler IJ. Selection of successive tumor lines for metastasis. Nature (New Biol) 1973; 242:148-149.
25. Fidler IJ. Biological behavior of malignant melanoma cells correlated to their survival in vivo. Cancer Res 1975; 35:218-224.
26. Hart IR. The selection and characterization of an invasive variant of the B16 melanoma. Am J Pathol 1979; 97:587-592.

27. Cohen JJ. Programmed cell death in the immune system. Adv Immunol 1991; 50:55-85.
28. Rao L, Debbas M, Sabbatini P, Hockenbery D, Korsmeyer S, White E. The adenovirus E1A proteins induce apoptosis, which is inhibited by the E1B 19 kDa and Bcl-2 proteins. Proc Natl Acad Sci USA 1992; 89:7742-7746.
29. Henderson S, Rowe M, Gregory C, Croom-Carter D, Wang F, Longnecker R, Kieff E, Rickinson A. Induction of bcl-2 expression by Epstein-Barr virus latent membrane protein 1 protects infected B cells from programmed cell death. Cell 1991; 65:1107-1115.
30. Wyllie AH. Glucocorticoid-induced thymocyte apoptosis is associated with endogenous endonuclease activation. Nature 1980; 284:555-556.
31. van Golen KL, Owen-Schaub L, Price JE. Metastatic and nonmetastatic K-1735 melanoma cells differ in Fas-mediated apoptosis and growth response to lung-conditioned medium. Proc Amer Assoc Cancer Res 1996; 37:79.
32. van Golen KL, Risin S, Staroselsky A, Berger D, Tainsky MA, Pathak S, Price JE. Predominance of the metastatic phenotype in hybrids formed by fusion of mouse and human melanoma clones. Clin Exp Metastasis 1996; 14:95-106.
33. Risin SA, van Golen KL, Price JE, Pathak S. Further evidence for involvement of human chromosome 17 in control of metastatic phenotype in melanoma. Proc Amer Assoc Cancer Res 1996; 37:79.
34. Bergman PJ, Kiefer JA, Price JE, Ley PB, Wynn P, Bucana C, OBrian CA, McConkey DJ. Acquisition of apoptosis resistance with chemoresistance and metastatic potential in human breast cancer: Role of bcl-2 and bcl-X_L. Proc Amer Assoc Cancer Res 1996; 37:7.
35. Glinsky GV. Cell adhesion and metastasis: Is the site specificity of cancer metastasis determined by leukocyte-endothelial cell recognition and adhesion? Crit Rev Oncol/Hemat 1993; 14:229-278.
36. Raz A, Zhu D, Hogan V, Shah N, Raz T, Karakash R, Pazerini G, Carmi P. Evidence for the role of 34 kDa galactoside-binding lectin in transformation and metastasis. Int J Cancer 1990; 46:871-877.
37. Raz A, Lotan R. Endogenous galactoside-binding lectins: A new class of functional tumor cell surface molecules related to metastasis. Cancer Metastasis Rev 1987; 6:433-452.
38. Lotan R, Raz A. Lectins in cancer cells. Ann NY Acad Sci 1988; 551:385-398.
39. Meromsky L, Lotan R, Raz A. Implications of tumor cell surface lectins as mediators of cellular interactions and lung colonization. Cancer Res 1986; 46:5270-5275.
40. Platt D, Raz A. Modulation of the lung colonization of B16-F1 melanoma cells by citrus pectin. J Natl Cancer Inst 1992; 84:438-439.
41. Schoeppner HL, Raz A, Ho SB, Bresalier RS. Expression of an endogenous galactose-binding lectins correlates with neoplastic progression in the colon. Cancer 1995; 75:2818-2826.

42. Bresalier RS, Yan P, Raz A. Expression of an endogenous galactose-binding lectin (galectin-3) correlates with malignancy of tumors in the CNS. Proc Amer Assoc Cancer Res 1996; 37:81.
43. Bresalier RS, Byrd JC, Raz A. Colon cancer mucin is a major ligand for the β-galactoside binding protein, galectin-3. Proc Amer Assoc Cancer Res 1996; 37:81.
44. Rice GE, Bevilacqua MP. An inducible endothelial cell surface glycoprotein mediates melanoma cell adhesion. Science 1989; 246:1303-1306.
45. Springer TA, Lasky LA. Sticky sugars for selectins. Nature 1991; 349:196-197.
46. Sawada R, Lowe JB, Fukuda M. E-selectin-dependent adhesion efficiency of colonic carcinoma cells is increased by genetic manipulation of their cell surface lysosomal membrane glycoprotein-1 expression levels. J Biol Chem 1993; 268:12675-12681.
47. Datta R, Manome Y, Taneja N, Boise LH, Weichselbaum R, Thompson CB, Slapak CA, Kute D. Overexpression of Bcl-X_L by cytotoxic drug exposure confers resistance to ionizing radiation-induced internucleosomal DNA fragmentation. Cell Growth & Differ 1995; 6:63-370.
48. Dole M, Nunez G, Merehant AK, Maybaum J, Rode CK, Bloch CA, Castle VP. Bcl-2 inhibits chemotherapy-induced apoptosis in neuroblastoma. Cancer Res 1994; 54:3253-3259.
49. McDonnell TJ, Meyn RF, Robertson LE. Implications of apoptotic cell death regulation in cancer therapy. Semin Cancer Biol 1995; 6:53-60.
50. Reed JC. Bcl-2: prevention of apoptosis as a mechanism of drug resistance. Hematol. Oncol. Clin N Am 1995; 9:451-474.
51. van Golen KL, Berger DH, Tainsky MA, Risin S, Price JE. Mutations at codons 249 and 273 of human p53 gene confer metastatic potential to nonmetastatic mouse melanoma cells. Proc Amer Assoc Cancer Res 1996; 37:79.
52. Lowe SW, Ruley HE, Jacks T, Housman DE. p53-dependent apoptosis modulates the cytotoxicity of anticancer agents. Cell 1993; 74:957-967.
53. Lowe SW, Bodis S, McClatchey A, Remington L, Ruley HE, Fisher DE, Housman DE, Jacks T. p53 status and the efficacy of cancer therapy in vivo. Science 1994; 266:807-810.
54. Symonds H, Krail I, Remington L, Saenz-Robles M, Lowe S, Jacks T, Van Dyke T. p53-dependent apoptosis supresses tumor growth progression in vivo. Cell 1994; 78:703-711.
55. Arends MJ, McGregor AH, Toft NJ, Brown EJH, Wyllie AH. Susceptibility to apoptosis is differentially regulated by c-myc and mutated Ha-ras oncogenes and is associated with endonuclease availability. Br J Cancer 1994; 68:1127-1133.
56. Wyllie AH. Apoptosis: the 1992 Frank Rose Memorial Lecture. Br J Cancer 1992; 67:205-208.

57. Wyllie AH, Rose KA, Morris RG, Steel CM, Foster E, Spandidos DA. Rodent fibroblast tumours expressing human myc and ras genes: growth, metastasis and endogenous oncogene expression. Br J Cancer 1987; 56:251-259.
58. Lin HJ, Eviner LV, Prendergast GC, White E. Activated H-ras rescues E1A-induced apoptosis and cooperates with E1A to overcome p53-dependent growth arrest. Mol Cell Biol 1995; 15:4536-4544.
59. Kinoshita T, Yokota T, Arai K, Miyajima A. Regulation of Bcl-2 expression by oncogenic Ras protein in hematopoetic cells. Oncogene 1993; 10:2207-2210.
60. Wyllie A. Apoptosis and the regulation of cell number in normal and neoplastic tissues: an overview. Cancer Metastasis Rev 1992; 11:95-103.
61. Sarraf CE, Bowen ID. Proportions of mitotic and apoptotic cells in a range of untreated experimental tumours. Cell Tissue Kinet 1988; 21:5-49.
62. Columbano A, Ledda-Columbano GM, Rao PM, Rajalakshmi S, Sarma DS. Occurence of cell death (apoptosis) in preneoplastic and neoplastic liver cells: a sequential study. Am J Pathol 1984; 116:441-446.
63. Aihara M, Truong LD, Dunn JK, Wheeler TM, Scardino PT, Thompson TC. Frequency of apoptotic bodies positively correlates with Gleason grade in prostate cancer. Hum Pathol 1994; 25:797-801.
64. Berges RR, Vukanovic J, Epstein JI, CarMichel M, Cisek L, Johnson DE, Veltri RW, Walsh PC, Isaacs JT. Implications of cell kinetic changes during the progression of human prostatic cancer. Clin Cancer Res 1995; 1:473-480.
65. Isacson C, Kessis TD, Hedrick L, Cho KR. Both cell proliferation and apoptosis increase with lesion grade in cervical neoplasia but do not correlate with human papillomavirus type. Cancer Res 1996; 56:669-674.
66. El-Labban NG, Osorio-Herrera E. Apoptotic bodies and abnormally dividing epithelial cells in squamous cell carcinoma. Histopathology 1986; 10:921-931.
67. Steel GG. Growth Kinetics of Tumors. London: Oxford University Press, 1977.
68. Deschner EE. Kinetics of Normal, Preneoplastic, and Neoplastic Colonic Epithelium. San Diego: Academic Press, Inc, 1990:41.
69. Bedi A, Pasricha PJ, Akhtar AJ, Barber JP, Bedi GC, Giardiello FM, Zehnbauer BA, Hamilton SR, Jones RJ. Inhibition of apoptosis during development of colorectal cancer. Cancer Res 1995; 55:1811-1816.
70. Krajewska M, Moss SF, Krajewski S, Song K, Holt PR, Reed JC. Elevated expression of Bcl-X and reduced Bak in primary colorectal adenocarcinomas. Cancer Res 1996; 56:2422-2427.
71. Vaux DL. Toward an understanding of the molecular mechanisms of physiological cell death. Proc Natl Acad Sci USA 1993; 90:786-789.
72. Oltvai ZN, Korsmeyer SJ. Checkpoints of dueling dimers foil death wishes. Cell 1994; 79:189-192.

73. Oltvai Z, Milliman C, Korsmeyer SJ. Bcl-2 heterodimerizes in vivo with a conserved homolog, Bax, that accelerates programmed cell death. Cell 1993; 74:609-619.
74. Boise LH, Gonzalez-Garcia M, Postema CE, Ding L, Lindsten T, Turka LA, Mao X, Nunez G, Thompson CB. *bcl-x*, a *bcl-2*-related gene that functions as a dominant regulator of apoptotic cell death. Cell 1993; 74:597-608.
75. Sato T, Hanada M, Bodrug S, Irie S, Iwama N, Boise LH, Thompson CB, Golemis E, Fong L, Wang H-G, Reed JC. Interactions among members of the bcl-2 protein family analyzed with a yeast two-hybrid system. Proc Natl Acad Sci USA 1994; 91:9238-9242.
76. Hanada M, Aime-Sempe C, Sato T, Reed JC. Structure-function analysis of bcl-2 protein: identification of conserved domains important for homodimerization with bcl-2 and heterodimerization with bax. J Biol Chem 1995; 270:11962-11968.
77. Yang E, Zha J, Jockel J, Boise LH, Thompson CB, Korsmeyer SJ. Bad: a heterodimeric partner for Bcl-X_L and Bcl-2, displaces bax and promotes cell death. Cell 1995; 80:285-291.
78. Farrow SN, White JHM, Martinou I, Raven T, Pun K-T, Grinham CJ, Martinou J-C, Brown R. Cloning of a bcl-2 homologue by interaction with adenovirus E1B 19K. Nature (Lond) 1995; 374:731-733.
79. Chittenden T, Harrington EA, O Connor R, Flemington C, Lutz RJ, Evan GI, Guild BC. Induction of apoptosis by the Bcl-2 homologue Bak. Nature (Lond) 1995; 374:733-736.
80. Kiefer MC, Brauer MJ, Powers VC, Wu JJ, Ubansky SR, Tomei LD, Barr PJ. Modulation of apoptosis by the widely distributed Bcl-2 homologue Bak. Nature (Lond) 1995; 374:736-739.
81. Bodrug SE, Aime-Sempe C, Sato T, Krajewski S, Hanada M, Reed JC. Biochemical and functional comparisons of Mcl-1 and Bcl-2 proteins: evidence for a novel mechanism of regulating Bcl-2 family protein function. Cell Death Differ 1995; 2:173-182.
82. Reynolds JE, Yang T, Qian L, Jenkinson JD, Zhou P, Eastman A, Craig RW. Mcl-1, a member of the Bcl-2 family, delays apoptosis induced by c-myc overexpression in Chinese chamster ovary cells. Cancer Res 1994; 54:6348-6352.
83. Chao DT, Linette GP, Boise LH, White LS, Thompson CB, Korsmeyer SJ. Bcl-X_L and Bcl-2 repress a common pathway of cell death. J Exp Med 1995; 182:821-828.
84. Silvetsrini R, Veneroni S, Daidone MG, Benini E, Boracchi P, Mezzetti M, Di Fronzo G, Rilke F, Veronesi U. The Bcl-2 protein: a prognostic indicator strongly related to p53 protein in lymph node-negative breast cancer patients. J Natl Cancer Inst 1994; 86:499-504.
85. Ealovega MW, McGinnis PK, Sumantran VN, Clarke MF, Wicha MS. *bcl-x_s* Gene therapy induces apoptosis of human mammary tumors in nude mice. Cancer Res 1996; 56:1965-1969.
86. Harris AL. Editorial: what does Bcl-2 mean in solid tumours friend or foe? Ann Oncol 1994; 5:388-390.

87. Krajewski S, Blomvqvist C, Franssila K, Krajewska M, Wasenius V-M, Niskanen E, Reed JC. Reduced expression of pro-apoptotic gene *Bax* is associated with poor response rates to combination chemotherapy and shorter survival in women with metastatic breast adenocarcinoma. Cancer Res 1995; 55:4471-4478.
88. Bargou RC, Daniel PT, Mapara MY, Bommert K, Wagener K, Wagener C, Kallinich B, Royer HD, Dorken B. Expression of the *bcl-2* gene family in normal and malignant breast tissues: low *bax-a* expression in tumor cells correlates with resistance towards apoptosis. Int J Cancer 1995; 60:854-859.
89. Miyashita T, Reed JC. Tumor suppressor p53 is a direct transcriptional activator of the human *bax* gene. Cell 1995; 80:293-299.
90. Sala A, Casella I, Grasso L, Bellon T, Reed JC, Miyashita T, Peschle C. Apoptotic response to oncogenic stimuli: cooperative and antagonistic interactions between c-myb and the growth suppressor p53. Cancer Res 1996; 56:1991-1996.
91. Hass-Kogan DA, Kogan SC, Levi D, Dazin P, T Ang A, Fung Y-KT, Israel MA. Inhibition of apoptosis by the retinoblastoma gene product. EMBO J 1995; 14:461-472.
92. Haupt Y, Rowan S, Oren M. p53-mediated apoptosis in Hela cells can be overcome by excess pRB. Oncogene 1995; 10:1563-1571.
93. An B, Dou QP. Cleavage of retinoblastoma protein during apoptosis: an interleukin 1β-converting enzyme-like protease as candidate. Cancer Res 1996; 56:438-442.
94. Foster R, Zhao X, Day K, Swanson P, Humphrey P, Dean D, Day M. The retinoblastoma gene product mediates apoptosis of prostate epithelium. Proc Amer Assoc Cancer Res 1996; 37:28.
95. Warburg O. Uber den stoffwechsel der tumoren. Cambridge University Press, 1926.
96. Sevick EM, Jain RK. Blood flow and venous pH of tissue-isolated Walker 256 carcinoma during hyperglycemia. Cancer Res 1988; 48:1202-1207.
97. Vaupel P, Kallinowski F, Okunieff P. Blood flow, oxygen and nutrient supply, and metabolic microenvironment. Cancer Res 1989; 49:6449-6465.
98. Stolk JA, Bhujwalla ZM, Blackband SJ, Shungu DC, Wehrle JP, Glickson JD. Quantitative In Vivo Measurement of Blood Flow in RIF-1 Tumors by $^{1}H/^{2}H$ NMR Spectroscopy. Proc 10th Ann Mtg Soc Magn Reson Med 1991:984.
99. Bhujwalla ZM, Tozer GM, Feld SB, Maxwell RJ, Griffiths JR. Energy metabolism of RIF-1 tumors following hydralazine. Radiother Oncol 1990; 19:281-291.
100. Wike-Hooley JL, Haveman J, Reinhold HS. The relevance of tumour pH to the treatment of malignant disease. Radiother Oncol 1991; 2:343-366.
101. Griffiths JR. Are cancer cells acidic? Br J Cancer 1991; 64:425-427.

102. Gillies RJ, Liu Z, Bhujwalla ZM. 31P NMR measurement of extracellular pH in situ using 3-aminopropylphosphonate. Am J Physiol 1994; 267(Cell Physiol 36):C195-C203.
103. Schlappack OK, Zimmermann A, Hill RP. Glucose starvation and acidosis: effect on experimental metastatic potential, DNA content, and MTX resistance of murine tumour cells. Int J Cancer 1991; 64:663-670.
104. Martinez-Zaguilen R, Seftor EA, Seftor REB, Chu Y-W, Gillies RJ, Hendrix MJC. Acidic pH enhances the invasive behavior of human melanoma cells. Clin Exp Metastasis 1996; 14:176-186.
105. Kato Y, Nakayama Y, Umeda M, Miyazaki K. Induction of 103 kDa gelatinase/type IV collagenase by acidic culture conditions in mouse metastatic melanoma cell lines. J Biol Chem 1992; 267:11424-11430.
106. Park H, Lyons J, Auger E, Song C. Apoptosis in cancer cells caused by low pH environment. Proc Amer Assoc Cancer Res 1996; 37:28.
107. Castle VP, Heidelberger JB, Ou X, Dole M, Nunez Z. Expression of the apoptosis-suppressing protein bcl-2, in neuroblastoma is associated with unfavorable histology and N-myc amplification. Am J Pathol 1993; 143:1543-1550.
108. McDonnell TJ, Troncoso P, Brisbay SM, Logothetis C, Chung LWK, Hsieh J-T, Tu S-M, Campbell ML. Expression of the protooncogene bcl-2 in the prostate and its association with emergence of androgen-independent prostate cancer. Cancer Res 1992; 52:6940-6944.
109. Hague A, Moorghen M, Hicks D, Chapman M, Paraskeva C. Bcl-2 expression in human colorectal adenomas and carcinomas. Oncogene 1994; 9:3367-3370.
110. Barry MA, Behnke CA, Eastman A. Activation of programmed cell death (apoptosis) by cisplatin, other anticancer drugs, toxins and hyperthermia. Biochem Pharmacol 1990; 40:2353-2362.
111. Eastman A. Activation of programmed cell death by anticancer agents: cisplatin as a model system. Cancer Cells 1990; 2:275-280.
112. Fisher DE. Apoptosis in cancer therapy: crossing the threshold. Cell 1994; 78:539-542.
113. Hickman JA. Apoptosis induced by anticancer drugs. Cancer Metastasis Rev 1992; 11:121-139.
114. Stewart BW. Mechanisms of apoptosis: integration of genetic, biochemical, and cellular indicators. J Natl Cancer Inst 1994; 86:1286-1296.
115. Minn AJ, Rudin CM, Boise LH, Thompson CB. Expression of Bcl-X_L can cofer a multidrug resistance phenotype. Blood 1995; 86:1903-1910.
116. Lotan R, Ito H, Yasui W, Yokozaki H, Lotan D, Tahara E. Expression of a 31 kDa lactose-binding lectin in normal gastric mucosa and in primary and metastatic gastric carcinomas. Int J Cancer 1994; 56:474-480.
117. Gehlsen KR, Davis GE, Sriramarao P. Integrin expression in human melanoma cells with differing invasive and metastatic properties. Clin & Exp Metastasis 1992; 10:111-120.

118. Albelda SM, Mette SA, Elder DE, Stewart R, Damjanovich L, Herlyn M, Buck CA. Integrin distribution in malignant melanoma: association of the β_3 subunit with tumor progression. Cancer Res 1990; 50:6757-6764.
119. McGregor BC, McGregor JL, Weiss LM, Wood GS, Hu CH, Boukerche H, Warnke RA. Presence of cytoadhesins (IIb- IIa-like glycoproteins) on human metastatic melanomas but not on benign melanocytes. Am J Clin Pathol 1989; 92:495-499.
120. Danen EH, Ten Berge PJ, van Muijen GN, van t Hoff-Grootenboer AE, Brocker EB, Ruiter DJ. Emergence of $\alpha_5\beta_1$ fibronectin- and $\alpha_v\beta_3$ vitronectin-receptor expression in melanocytic tumour progression. Histopathology (Oxford) 1994; 24:249-256.
121. Cheresh DA. The biological function of β_3 integrins and other vitronectin receptors. In: Metcalf BW, Dalton BJ, Poste G, eds. Cellular Adhesion: Molecular Definition to Therapeutic Potential. New York: Plenum Press, 1994:3-24.
122. Leavesley DI, Ferguson GD, Wayner EA, Cheresh DA. Requirement of the integrin β_3 subunit for carcinoma cell spreading or migration on vitronectin and fibrinogen. J Cell Biol 1992; 117:1101-1107.
123. Yum Z, Menter DG, Nicolson GL. Involvement of integrin $\alpha_v\beta_3$ in cell adhesion, motility, and liver metastasis of murine RAW117 large cell lymphoma. Cancer Res 1996; 56:3103-3111.
124. Shibata M-A, Maroulakou I, Jorcyk CL, Gold LG, Ward JM, Green JE. p53-independent apoptosis during mammary tumor progression in C3(1)/SV40 large T antigen transgenic mice: suppression of apoptosis during the transition from preneoplasia to carcinoma. Cancer Res 1996; 56:2998-3003.
125. Frisch SM, Vuori K, Ruoslahti E, Chan-Hui, P-Y. Control of adhesion-dependent cell survival by focal adhesion kinase. J Cell Biol 1996; 134:793-799.
126. Owens LV, Xu L, Craven RJ, Dent GA, Weiner TM, Kornberg L, Liu ET, Cance WG. Overexpression of the focal adhesion kinase $p125^{FAK}$) in invasive human tumors. Cancer Res 1996; 55:2752-2755.
127. Beg AA, Baltimore D. An essential role for NF-κB in preventing TNF-α-induced cell death. Science 1996; 274:782-784.
128. Wang C-Y, Mayo MW, Baldwin AS. TNF- and cancer therapy-induced apoptosis potentiation by inhibition of NF-κB. Science 1996; 274:784-787.
129. Van Antwerp DJ, Martin SJ, Kafri T, Green DR, Verma IM. Suppression of TNF-α-induced apoptosis by NF-κB. Science 1996; 274:787-789.
130. Brockman JA, Scherer DC, McKinsey TA, Hall SM, Qi X, Lee WY, Ballard DW. Coupling of a signal response domain in IκBα to multiple pathways for NF-κB activation. Mol Cell Biol 1995; 15:2809-2818.

131. Brown K, Gerstberger S, Carlson L, Franzoso G, Siebenlist U. Control of IkB-α proteolysis by site-specific, signal-induced phosphorylation. Science 1995; 267:1485-1488.
132. DiDonato J, Mercurio F, Rosette C, Wu-Li J, Suyang H, Ghosh S, Karin M. Mapping of the inducible IκB phosphorylation sites that signal its ubiquitination and degradation. Mol Cell Biol 1996; 16:1295-1304.
133. Palombella VJ, Rando OJ, Goldberg AL, Maniatis T. The ubiquitin-proteasome pathway is required for processing the NF-κB1 precursor protein and the activation of NF-κB. Cell 1994; 78:773-785.
134. Kobayashi H, Man S, Graham CH, Kapitain SJ, Teicher BA, Kerbel RS. Acquired multicellular-mediated resistance to alkylating agents in cancer. Proc Natl Acad Sci USA 1993; 90:3294-3298.
135. Graham CH, Kobayashi H, Stankiewicz KS, Man S, Kapitain SJ, Kerbel RS. Rapid acquisition of multicellular drug resistance after a single exposure of mammary tumor cells to antitumor alkylating agents. J Natl Cancer Inst 1994; 86:975-982.
136. Kerbel RS, Rak J, Kobayashi H et al. Multicellular resistance: a new paradigm to explain aspects of acquired drug resistance of solid tumors. Cold Spring Harbor Symposium on Quantitative Biology: Molecular Genetics of Cancer, 1994; 59:661-672.
137. Kerbel RS, St Croix B, Rak J et al. Is there a role for "anti-adhesives" as chemosensitizers in the treatment of solid tumors by chemotherapy? Bulletin l'Institut Pasteur 1995; 92:248-256.
138. St Croix B, Rak JW, Kapitain S, Sheehan C, Graham CH, Kerbel RS. Hyaluronidase reverses adhesion-dependent multicellular drug resistance in mammary carcinoma cells. J Natl Cancer Inst 1996; 88:1285-1296.
139. Frankel A, Glinsky GV, Mossine VV, Buckman R, Kerbel RS. Synthetic glycoamine analogs synergize with taxol and cisplatin in inducing programmed cell death in ovarian cancer cells. Proc Amer Assoc Cancer Res 1996; 38:505.
140. Glinsky GV, Glinsky VV, Ivanova AB, Hueser CN. Apoptosis and metastasis: increased apoptosis resistance of metastatic cancer cells is associated with the profound deficiency of apoptosis execution mechanisms. Cancer Lett 1996; 115:185-193.
141. McConkey DJ, Greene G, Pettaway CA. Apoptosis resistance increases with metastatic potential in cells of the human LNCap prostate carcinoma line. Cancer Res 1996; 56:5594-5599.
142. Yang RY, Hsu DK, Liu FT. Expression of galectin-3 modulates T-cell growth and apoptosis. Proc Natl Acad Sci USA 1996; 93:6737-6742.
143. Glinsky GV. Apoptosis in metastatic cancer cells. Crit Rev Oncol/Hemat 1997; 25:175-186.

144. Shimizu S et al. Induction of apoptosis as well as necrosis by hypoxia and predominant prevention of apoptosis by Bcl-2 and Bcl-X_L. Cancer Res 1966; 56:2161-2166.
145. Graeber TG et al. Hypoxia-mediated selection of cells with diminished apoptotic potential in solid tumors. Nature 1996; 379:88-91.
146. Helmlinger G, Yan F, Dellian M, Jain RK. Interstitial pH and pO_2 gradients in solid tumours in vivo: High resolution measurements reveal a lack of correlation. Nature Medicine 1997; 3:177-182.
147. Gajewski TF, Thompson CB. Apoptosis meets signal transduction: elimination of a BAD influence. Cell 1996; 87:589-592.

CHAPTER 3

Apoptosis Resistance of Metastatic Cancer Cells Is Associated with Deficiency of Apoptosis Execution Mechanisms

3.1. Summary

Programmed cell death, particularly adhesion-dependent regulation of cell survival and apoptosis, is recognized as one of the main homeostatic mechanisms designed to control cell positioning, eliminate misplaced cells and block metastatic dissemination. Recently we reported that highly metastatic cancer cells exhibit a higher resistance to programmed cell death compared to their poorly metastatic counterparts.[11] However, the molecular and genetic basis for the association of aggressive metastatic phenotype with resistance toward apoptosis remains to be elucidated. Here we extended our investigation on apoptosis and metastasis using a panel of nine murine and human cancer cell lines with different metastatic potential. We examined the relationship of the metastatic ability and the sensitivity to apoptosis as well as determined the status of two major apoptosis execution mechanisms (induction of nuclear Ca^{2+}-dependent endonucleases and activation of ICE-like proteases) in cancer cells with distinct metastatic potential and different sensitivity to apoptosis. We found that high metastatic potential is strictly associated with increased resistance to apoptosis, diminished level of nuclear Ca^{2+}-dependent endonucleases, and significantly reduced activity of CPP32/Yama death protease. We concluded that

Failure of Apoptosis and Cancer Metastasis, by Gennadi V. Glinsky.

high resistance to apoptosis of metastatic cancer cells is associated with and may depend upon the profound deficiency of major apoptosis execution mechanisms.

3.2. Introduction

Programmed cell death (apoptosis) is considered an important homeostatic mechanism that balances cell production with cell death and maintains correct cell numbers in the body in physiological and pathological conditions.[1-4] Aberration in regulation of apoptosis is expected to contribute to tumor progression and oncogenesis,[3-5] and cancer cells would gain a selective growth advantage by blocking the apoptosis.[3,4,6] The recent findings on adhesion-dependent regulation of programmed cell death suggest that apoptosis may also control cell positioning in the body through proper cell-cell and cell-extracellular matrix interactions[7-10] (reviewed in ref. 32). Matrix-independent passage of cancer cells through blood and/or lymph compartments is an essential component of the metastatic cascade. Therefore, metastatic cancer cells by definition would have to acquire the ability to survive in a detached state. Furthermore, metastatic dissemination may depend upon the resistance of metastatic cells to apoptosis, since detachment of cells from the extracellular matrix induces programmed cell death[7-10] and malignant transformation confers at least partial resistance to apoptosis induced by disruption of cell-matrix interactions.[7,8] These data support the hypothesis that apoptosis is one of the key mechanisms for elimination of misplaced cells and may prevent metastatic dissemination.[2,4] However, there is little direct experimental evidence linking programmed cell death and metastasis. We studied the responses to different external signals leading to apoptosis of several poorly and highly metastatic cell lines employing the murine B16 melanoma experimental metastasis model.[11] We found that highly metastatic cells exhibit a superior survival ability and resistance to apoptosis compared to poorly metastatic cells. We also showed that the survival of metastatic cancer cells is associated with their enhanced homotypic aggregation. Thus, our results indicate that there is a genetic and molecular basis linking the aggressive metastatic phenotype with resistance to apoptosis.

Superior resistance to apoptosis would give the metastatic cancer cells an obvious selective growth advantage during tumor progression.[32] Moreover, elucidation of the molecular mechanisms of resistance to apoptosis by metastatic cancer cells may help in discovery of novel therapeutic targets for antimetastatic drug design. It has been shown recently that synthetic low molecular weight inducers of apoptosis in metastatic cancer cells act as effective inhibitors of human breast cancer metastasis in nude mice.[12,13]

Many apoptosis induction and initiation pathways in mammalian cells result in common apoptosis execution mechanisms destroying the key nucleic acid and protein targets. One of the main biochemical features that distinguish programmed cell death is the induction of the nuclear Ca^{2+}-dependent endonucleases and DNA fragmentation in intranucleosomal regions (characteristic ladder pattern of DNA fragmentation).[14-16] Experimental evidence has also suggested a high degree of conservation of the ICE-like protease-dependent pathway of apoptosis among species and various initiators of programmed cell death.[17-21] Recently a new nomenclature has been recommended for the human apoptosis-associated ICE/CED-3 death protease family which proposes the common name "caspase" as a root for serial names for all family members.[35] Here we will describe the experiments that extended our investigation on apoptosis and metastasis using a panel of nine murine and human cancer cell lines with different metastatic potentials. We examined the relationship of the metastatic ability and the sensitivity to saturation density- and serum deprivation-induced apoptosis as well as determined the status of two major apoptosis execution mechanisms (induction of nuclear Ca^{2+}-dependent endonucleases and activation of ICE-like proteases) in cancer cells with distinct metastatic potential and different sensitivity to apoptosis. We found that high metastatic potential is strictly associated with increased resistance to apoptosis, diminished level of nuclear Ca^{2+}-dependent endonucleases, and significantly reduced activity of CPP32/Yama death protease (caspase 3). We concluded that high resistance to apoptosis of metastatic cancer cells is associated with and may depend upon the profound deficiency of major apoptosis execution mechanisms, targeted at the key nucleic acid and protein death substrates.

3.3. Association of High Metastatic Potential with Increased Resistance Toward Apoptosis

Employing a murine B16 melanoma experimental metastasis model,[26-29] we reported earlier that highly metastatic cancer cells exhibit a superior survival ability and increased resistance to apoptosis compared to poorly metastatic counterparts.[11] Here we extend our observation employing six human cancer cell lines with established striking differences in in vivo metastatic propensity.[12,13,22-25] Highly metastatic human breast carcinoma MDA-MB-435 cells showed significantly lower levels of apoptosis both in saturation density and serum deprivation experiments compared to MDA-MB-231 cells with intermediate metastatic potential, as well as compared to poorly metastatic MDA-MB-468 and MDA-MB-361 cells (Fig. 3.1A,G). Similarly, highly metastatic human melanoma TXM-13 cells were ~10-fold less sensitive to apoptosis compared to poorly metastatic DX-3 cells (Fig. 3.1B). The difference in apoptosis levels between poorly and highly metastatic cancer cells as determined by the TUNEL assay was in good agreement with the DNA fragmentation analysis (Fig. 3.1D,E,F), indicating that substantially higher levels of DNA degradation and accumulation of apoptotic DNA fragments occurred in poorly metastatic cancer cells.

Thus, the highly aggressive metastatic phenotype is strictly associated with an increased resistance toward apoptosis in both murine and human cancer cells. Similar experimental observation was recently published employing human prostate cancer cell lines with differing metastatic potential.[30]

3.4. Activation of ICE-like Protease(s) is Required for Execution of Serum Deprivation-Induced Apoptosis

Experimental evidence has suggested a high degree of conservation of the ICE-like protease-dependent pathway of apoptosis among species and various initiators of programmed cell death.[17-21] To determine whether the activation of ICE-like proteases is required for execution of serum deprivation-induced apoptosis, we next examined the effect of the specific cell permeable ICE-like protease inhibitor, peptide N-benzyloxycarbonyl-Val-Ala-Asp-fluoromethylketone (Z-VAD-fmk), on serum starvation-induced apoptosis in MDA-MB-435 cells. Treatment of the MDA-MB-435 human breast carcinoma cells with 50 μM concentration of the ICE-like protease inhibitor Z-VAD-fmk almost completely blocked the serum deprivation-induced apoptosis, as determined by the TUNEL assay (Fig.

3.2). Thus, functional ICE-like protease (or proteases) is required for execution of serum deprivation-induced apoptosis in metastatic human breast carcinoma cells. Correspondingly, the differential sensitivity to apoptosis of cancer cell lines with different metastatic potential may depend upon a difference in activity of ICE-like proteases.

3.5. Low Level of Apoptosis in Highly Metastatic Cancer Cells Is Correlated with Reduced Activity of the CPP32/Yama Death Protease

Among the broad family of apoptosis-associated ICE-like proteases in mammalian cells, the CPP32/Yama death protease is considered one of the key enzymes largely responsible for degradation of several critical nuclear protein targets during apoptosis.[17-20] The cleavage of the nuclear death protein substrates such as poly (ADP-ribose) polymerase (PARP) and the 70 kDa protein component of the U1 small nuclear ribonucleoprotein (U1-70kDa) by the CPP32/Yama may play a central role in apoptosis execution since it will terminate DNA repair and prevent mRNA splicing.[19,21] We next assessed the activity of the CPP32/Yama death protease in cancer cells with different metastatic potential employing a specific peptide substrate for CPP32/Yama, synthetic peptide Ac-DEVD-pNA (Calbiochem). To eliminate the potential interference of nonspecific hydrolysis due to contaminating proteinases,[31] the assay was performed using 100 μM of the specific peptide inhibitor of CPP32/Yama, peptide Ac-DEVD-CHO. As shown in Figure 3.3, the activity of the CPP32/Yama was significantly lower (two- to four-fold) in highly metastatic cancer cells compared to their corresponding poorly metastatic counterparts. Furthermore, the induction of apoptosis by serum deprivation in poorly metastatic B16-F1 melanoma cells was accompanied by substantial elevation in the activity of CPP32/Yama which was not detected in highly metastatic B16-BL6 melanoma cells (Fig. 3.3b). Thus, highly metastatic cancer cells exhibit significantly reduced activity of the CPP32/Yama death protease compared to the poorly metastatic counterparts. The low activity of the CPP32/Yama death protease may contribute to the increased resistance of highly metastatic cancer cells to apoptosis.

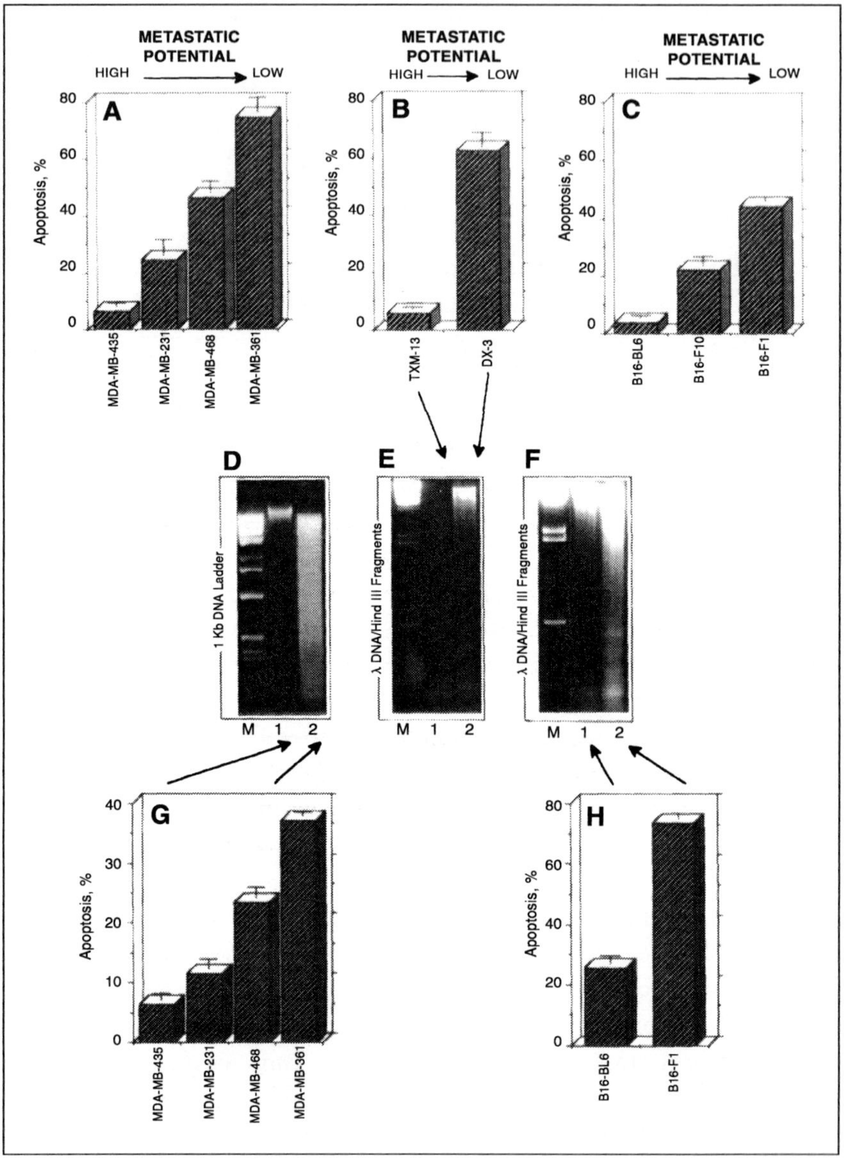

Fig. 3.1. Increased apoptosis resistance of human and murine highly metastatic cancer cells. Apoptosis level as determined by the TUNEL assay (top and bottom panels) or DNA fragmentation analysis (middle panel) in highly (B16-BL6 and B16-F10) and poorly (B16-F1) metastatic murine melanoma cell variants; in a panel of human breast carcinoma cell lines exhibiting high (MDA-MB-435), intermediate (MDA-MB-231), and low (MDA-MB-361 MDA-MB-468) metastatic potential in nude mice;[22,23] and in human melanoma cell lines highly (TXM-13) and poorly (DX-3) metastatic in nude mice.[24,25] Cell viability in medium containing either 10% or 0.1% FCS was assayed by acridine orange staining and/or trypan blue exclusion as previously described.[11] Identification of apoptotic cells employing the TUNEL method (terminal deoxynucleotidyl trans

3.6. Increased Resistance toward Apoptosis in Highly Metastatic Cancer Cells Is Associated with Low Activity of Nuclear Ca^{2+}-Dependent Endonucleases

Activation of nuclear Ca^{2+}-dependent endonucleases is responsible for the DNA fragmentation during apoptosis.[14-16] Moreover, increased levels of apoptosis in poorly metastatic cancer cells are correlated with higher levels of DNA fragmentation and apoptotic DNA accumulation (Fig. 3.1), suggesting enhanced activity of apoptosis-associated endonucleases. We therefore investigated whether Ca^{2+}-dependent endonuclease activity in isolated nuclei from murine and human melanoma and breast carcinoma cells is correlated with the apoptosis level and metastatic potential. High levels of apoptosis in poorly metastatic B16-F1 and DX-3 melanoma cells and MDA-MB-468 breast carcinoma cells were associated with the high levels of endogenous endonucleases available for direct activation by Ca^{2+} in isolated nuclei (Fig. 3.4). In contrast, highly metastatic MDA-MB-435 breast carcinoma cells as well as B16-BL6 and TXM-13 melanoma cells that exhibit low levels of apoptosis also express little to none of the nuclear Ca^{2+}-dependent endonuclease activity (Fig. 3.4). Thus, the levels of apoptosis in a panel of cancer cell lines with different metastatic potential are inversely correlated with the activity of nuclear Ca^{2+}-dependent endonucleases and metastatic ability, suggesting that nuclear endonuclease deficiency may play a role in resistance to apoptosis by metastatic cancer cells.

Fig. 3.2 Cont.
ferase-mediated dUTP-X nick end labeling method) was performed at day 2 (serum deprivation experiments; bottom panel) or day 6 (saturation density experiments; top panel) in each culture exactly as reported earlier.[11] The TUNEL assay in saturation density experiments with murine melanoma cells (C) was performed at day 4. The In Situ Cell Death Detection Kit, POD (Boehringer Mannheim Co., Indianapolis, IN) was employed, using fluorescein-dUTP and an anti-fluorescein antibody conjugated to a peroxidase reporter molecule as detection reagent. DNA fragmentation analysis (middle panel) was performed at day 2 (both in serum deprivation and saturation density experiments) in each culture exactly as described.[11] Cell culture conditions in both serum deprivation (0.1% FCS) and saturation density (medium containing 10% FCS) experiments were identical to those previously reported.[11] The panels of murine melanoma cell variants, human melanoma cells, and human breast carcinoma cell lines exhibiting different metastatic potential were previously described.[11,12,22-29] Percent of apoptotic cells is plotted. The results shown are representative of at least three separate experiments. Reproduced with permission from Glinsky GV et al, Cancer Lett 1996; 115:185-193.

3.7. Conclusion: Deficiency of the Two Major Apoptosis Execution Mechanisms as a Potential Molecular Cause of Increased Apoptosis Resistance in Metastatic Cancer Cells

In this chapter we described our experiments employing nine murine and human cancer cell lines with different metastatic potentials[34] that further extend our initial observation[11] that highly aggressive metastatic phenotype is strictly associated with increased resistance toward apoptosis. We provide experimental evidence that resistance to apoptosis of highly metastatic cancer cells is associated with the profound deficiency of the two major apoptosis execution mechanisms responsible for destruction during apoptosis of the main nucleic acid and protein death substrates. We determined

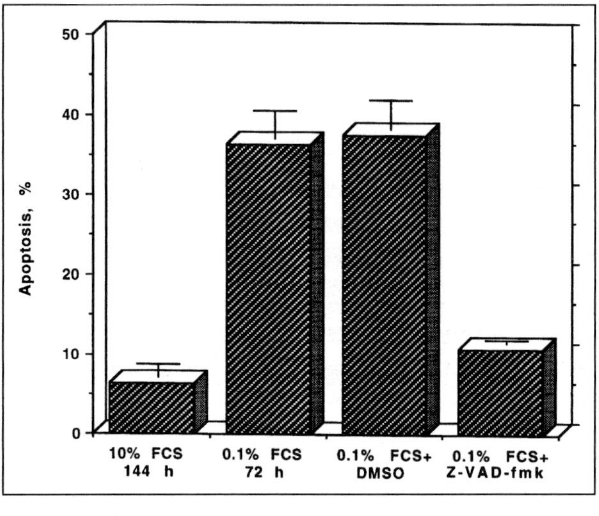

Fig. 3.2. Effect of specific ICE-like protease inhibitor peptide N-benzyloxycarbonyl-Val-Ala-Asp-fluoromethylketone (Z-VAD-fmk) on serum deprivation-induced apoptosis in MDA-MB-435 cells.
Cells were harvested from subconfluent cultures (60-70% confluence), plated at 4×10^5 ml^{-1} in multiple flasks with and without the specific peptide inhibitor of ICE-like protease Z-VAD-fmk (50 μM; Enzyme System Products, Dublin, CA), and cultivated for 3 days in medium containing 0.1% FCS. Cell viability was assayed by acridine orange staining and/or trypan blue exclusion as previously described.[11] Identification of apoptotic cells employing the TUNEL method (terminal deoxynucleotidyl transferase-mediated dUTP-X nick end labeling method) was performed exactly as reported earlier.[11] Cell culture conditions in serum deprivation experiments (0.1% FCS) were identical to those previously reported.[11] Percent of apoptotic cells is plotted. For every experimental condition, at least two separate experiments were conducted. Reproduced with permission from Glinsky GV et al, Cancer Lett 1996; 115:185-193.

that high metastatic potential is associated with diminished levels of nuclear Ca^{2+}-dependent endonucleases and with significantly reduced activity of CPP32/Yama death protease (caspase 3). It would be of interest to investigate whether the profound deficiency of apoptosis execution mechanisms in highly metastatic cancer cells is a consequence of upstream regulatory aberrations or reflection of a defect in apoptosis execution program per se. Using a panel of established cancer cell lines with different metastatic potential, phenotypic association of high metastatic potential with increased resistance toward apoptosis was clearly demonstrated for several rodent and human metastatic models such as murine B16 melanoma experimental metastasis model,[11] human prostate cancer cells,[30] human breast carcinoma and melanoma cell lines.[34] Gene transfer ex-

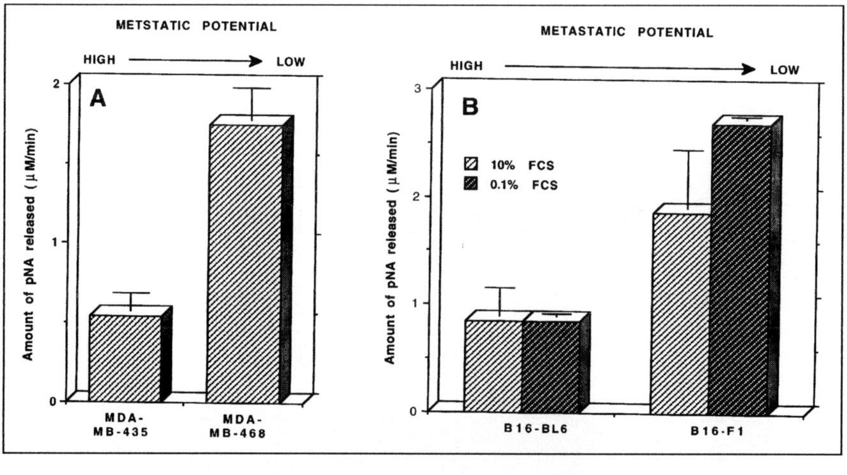

Fig. 3.3. The level of CPP32/Yama death protease is significantly reduced in cancer cells with higher resistance to apoptosis and high metastatic potential.
Human breast carcinoma cells (A) were grown to 60-70% confluence in medium containing 10% FCS, then cells were harvested and plated in medium containing 10% FCS as described,[11,12] cultured for 48 hours, harvested and CPP32/Yama assay was performed. Murine melanoma cells (B) were grown to 60-70% confluence in medium containing 10% FCS, then fresh cell culture medium was added containing either 10% or 0.1% FCS as indicated. Cells were cultured for an additional 48 hours, harvested and CPP32/Yama assay was performed as described. The representative results of one of two separate experiments are shown. Reproduced with permission from Glinsky GV et al, Cancer Lett 1996; 115:185-193.

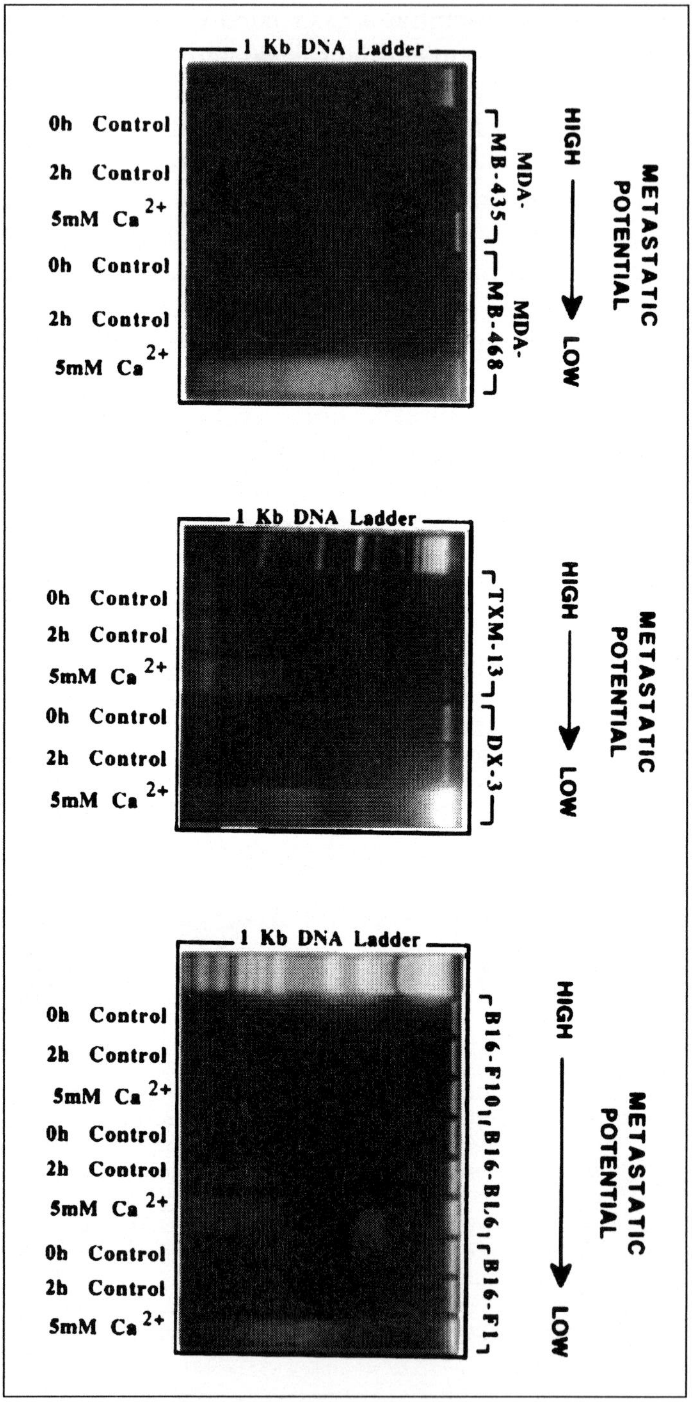
1 Kb DNA Ladder
0h Control
2h Control
5mM Ca 2+
0h Control
2h Control
5mM Ca 2+
MDA-MB-435
MDA-MB-468
METASTATIC POTENTIAL
HIGH
LOW
1 Kb DNA Ladder
0h Control
2h Control
5mM Ca 2+
0h Control
2h Control
5mM Ca 2+
TXM-13
DX-3
METASTATIC POTENTIAL
HIGH
LOW
1 Kb DNA Ladder
0h Control
2h Control
5mM Ca 2+
0h Control
2h Control
5mM Ca 2+
0h Control
2h Control
5mM Ca 2+
B16-F10
B16-BL6
B16-F1
METASTATIC POTENTIAL
HIGH
LOW

periments employing above referenced metastasis models would be essential to establish a genetic association of superior resistance toward apoptosis with aggressive metastatic phenotype. It will be interesting to determine whether translocation mechanisms of activated caspases to nuclei is altered in metastatic cancer cells and whether accessibility of the key nucleic acid and protein death substrates is modified during apoptosis execution in cancer cells.

References

1. Wyllie AH, Kerr JFR, Currie AR. Cell death: The significance of apoptosis. Int Rev Cytol 1980; 68:251-306.
2. Thompson CB. Apoptosis in the pathogenesis and treatment of disease. Science 1995; 267:1456-1462.
3. Williams GT, Smith CA. Molecular regulation on apoptosis: Genetic control on cell death. Cell 1993; 74:777-779.
4. Raff MC. Social control on cell survival and cell death. Nature 1992; 356:397-400.
5. Williams GT. Programmed cell death: Apoptosis and oncogenesis. Cell 1991; 65:1097-1098.
6. Reed JC. Bcl-2 and the regulation of programmed cell death. J Cell Biol 1994; 124:1-6.
7. Ruoslahti E, Reed JC. Anchorage dependence, integrins, and apoptosis. Cell 1994; 77:477-478.
8. Frisch SM, Francis H. Disruption of epithelial cell-matrix interactions induces apoptosis. J Cell Biol 1994; 124:619-626.
9. Meredith JE, Fazeli B, Schwartz MA. The extracellular matrix as a cell survival factor. Mol Biol Cell 1993; 4:953-961.
10. Boudreau N, Sympson CJ, Werb Z, Bissell MJ. Suppression of ICE and apoptosis in mammary epithelial cells by extracellular matrix. Science 1995; 267:891-893.
11. Glinsky GV, Glinsky VV. Apoptosis and metastasis: A superior resistance of metastatic cancer cells to programmed cell eath. Cancer Letters 1996; 101:43-51.

Fig. 3.4. The levels of endogenous nuclear Ca^{2+}-dependent endonucleases available for direct in vitro activation by calcium in isolated nuclei are diminished in cancer cells with higher resistance to apoptosis and high metastatic potential.
Cells were harvested from subconfluent cultures (60-70% confluence), plated at 4×10^5 ml^{-1} in multiple flasks and cultivated in medium containing 10% FCS. Cell viability was assayed by acridine orange staining and/or trypan blue exclusion as previously described.[11,12] Nuclei were isolated from subconfluent cultures (60-70% confluence) as described[33] and incubated for 2 hours in TBS in the absence or presence of 5 mM Ca^{2+}. DNA fragmentation in low molecular weight DNA extracts from identical number of control and calcium-treated nuclei was determined by agarose gel electrophoresis as previously described.[11,12] For every experimental condition, at least two separate experiments were conducted. Reproduced with permission from Glinsky GV et al, Cancer Lett 1996; 115:185-193.

12. Glinsky GV, Price JE, Glinsky VV, Mossine VV, Kiriakova G, Metcalf JB. Inhibition of human breast cancer metastasis in nude mice by synthetic glycoamines. Cancer Res 1996; 56:5319-5324.
13. Glinsky GV, Mossine VV, Price JE, Bielenberg D, Glinsky VV, Ananthaswamy HN, Feather MS. Inhibition of colony formation in agarose of metastatic human breast carcinoma and melanoma cells by synthetic glycoamine analogs. Clin & Exper. Metastasis 1996; 14:253-267.
14. Wyllie AH. Glucocorticoid-induced thymocyte apoptosis is associated with endogenous endonuclease activation. Nature 1980; 284:55-556.
15. Arends MJ, McGregor AH, Toft NJ, Brown EJH, Wyllie AH. Susceptibility to apoptosis is differntially regulated by c-myc and mutated Ha-ras oncogenes and is associated with endonuclease availability. Br J Cancer 1993; 68:1127-1133.
16. McConkey DJ, Chandra J, Wright S, Plunkett W, McDonnell TJ, Reed JC, Keating MJ. Apoptosis sensitivity in chronic lymphocytic leukemia is determined by endogenous endonuclease content and relative expression of Bcl-2 and Bax. J Immunol 1996; 156:2624-2630.
17. Fernandes-Alnemri T, Litwack G, Alnemri ES. CPP32, a novel human apoptotic protein with homology to Caenorhabditis elegans cell death protein Ced-3 and mammalian interleukin-1β-converting enzyme. J Biol Chem 1994; 269:30761-30764.
18. Nicholson DW, Ali A, Thornberry NA, Vaillancourt JP, Ding CK, Gallant M et al. Identification and inhibition of the ICE/CED-3 protease necessary for mammalian apoptosis. Nature 1995; 376:37-39.
19. Tewari M, Quan LT, O'Rourke K, Desnoyers S, Zeng Z, Beidler DR, Poirier GG, Salvesen GS, Dixit VM. Yama/CPP32b, a mammalian homolog of CED-3, is a crm-inhibitable protease that cleaves the death substrate poly (ADP-ribose) polymerase. Cell 1995; 81:801-809.
20. Lazebnik YA, Kaufmann SH, Desnoyers S, Poirier GG, Earnshaw WC. Cleavage of poly (ADP-ribose) polymerase by a proteinase with properties like ICE. Nature 1994; 371:346-347.
21. Casciola-Rosen LA, Miller DK, Anhalt GJ, Rosen A. Specific cleavage of the 70 kDa protein component of small nuclear ribonucleoprotein is a characteristic biochemical feature of apoptotic cell death. J Biol Chem 1994; 269:30757-30760.
22. Zhang RD, Fidler IJ, Price JE. Relative malignant potential of human breast carcinoma cell lines established from pleural effusions and a brain metastasis. Invasion & Metastasis 1991; 11:204-215.
23. Price JE, Polyzos A, Zhang RD, Daniels LM. Tumorigenicity and metastasis of human breast carcinoma cell lines in nude mice. Cancer Res 1990; 50:717-721.
24. Schackert G, Price JE, Zhang R, Bucana CD, Itoh K, Fidler IJ. Regional growth of different human melanomas as metastases in the brain of nude mice. Am J Pathol 1990; 136:95-102.

25. Zhang RD, Price JE, Schackert G, Itoh K, Fidler IJ. Malignant potential of cells isolated from lymph node or brain metastases of melanoma patients and implication for prognosis. Cancer Res 1991; 51:2029-2035.
26. Fidler IJ. The relationship of embolic homogeneity, number, size and viability to the incidence of experimental metastasis. Eur J Cancer 1973; 9:223-227.
27. Fidler IJ. Selection of successive tumor lines for metastasis. Nat New Biol 1973; 242:148-149.
28. Fidler IJ. Biological behavior of malignant melanoma cells correlated to their survival in vivo. Cancer Res 1975; 35:218-224.
29. Hart IR. The selection and characterization of an invasive variant of the B16 melanoma. Am J Pathol 1979; 97:587-592.
30. McConkey DJ, Greene G, Pettaway CA. Apoptosis resistance increases with metastatic potential in cells of the human LNCaP prostate carcinoma line. Cancer Res 1996; 6:5594-5599.
31. Thornberry NA. Interleukin-1β-converting enzyme. Methods Enzymology 1994; 244:615-631.
32. Glinsky GV. Apoptosis in metastatic cancer cells. Crit Rev Oncol/Hemat 1997; 25:175-186.
33. Marzluff WF, Huang RCC. Transcription and translation: a practical approach. In: Hames BD, Higgins SJ, eds. OxfordL IRL Press, 1985:89-129.
34. Glinsky GV, Glinsky VV, Ivanova AB, Hueser CN. Apoptosis and metastasis: increased apoptosis resistance of metastatic cancer cells is associated with the profound deficiency of apoptosis execution mechanisms. Cancer Lett 1997; 115:185-193.
35. Alnemri ES, Livingston DJ, Nicholson DW, Salvesen G, Thornberry N, Wong WW, Yuan J. Human ICE/CED-3 protease nomenclature. Cell 1996; 87:171.

CHAPTER 4

A Hypothesis for Telomere-Dependent Control of Apoptosis*

4.1. Summary

Telomere-dependent replicative cell senescence is one of the fundamental mechanisms determining proliferative potential, survival and aging of eukaryotic cells. Apoptosis is an important homeostatic mechanism that balances correct cell number and positioning in the body. No experimental evidence has been previously presented that may link these two phenomena. In this chapter we present three independent lines of evidence strongly indicating that telomere status may control sensitivity to apoptosis in metastatic cancer cells.

1. Induction of apoptosis and increased activity of nuclear Ca^{2+}-dependent endonucleases were associated with the enhanced asymmetrical degradation of the C-rich strand of telomeric DNA, suggesting activation of a specific telomere fragmentation mechanism during apoptosis.
2. Treatment of MDA-MB-435 human breast carcinoma cells with the short telomere-specific oligonucleotide $(TTAGGG)_3$ significantly inhibited serum deprivation-induced apoptosis with no interference with CPP32/Yama death protease activation, implying that activation of the CPP32/Yama required for apoptosis execution was not sufficient when cells were treated with TTAGGG telomere-specific oligonucleotides.

* *This chapter was written in coauthorship with Dr. Vladislav V. Glinsky, M.D.*

Failure of Apoptosis and Cancer Metastasis, by Gennadi V. Glinsky.

3. The cells with stable experimentally elongated telomeres manifested a substantially higher resistance to apoptosis, decreased activity of ICE-like death proteases (caspases) and diminished inducibility of Ca^{2+}-dependent nuclear endonucleases compared to the parental cells, confirming the significance of telomere length in regulation of apoptosis.

We employed the PCR-based assay with telomere-specific primers for analysis of telomeric repeat degradation products (TRDP) in low molecular weight DNA extracts from eight murine and human cancer cell lines. The TRDP assay revealed an asymmetrical accumulation of $(CCCTAA)n$ telomere fragments in proliferating in vitro melanoma and breast carcinoma cells, suggesting a nuclease-mediated G-tail formation mechanism on chromosome ends. We document that enhanced degradation of telomeric DNA is associated with a higher level of apoptosis as well as correlated with the induction of apoptosis in a panel of murine and human cancer cell lines exhibiting different metastatic potential and different apoptosis level. Decreased degradation of telomeric DNA and diminished levels of apoptosis were associated with high metastatic potential. Serum deprivation-induced apoptosis in metastatic cancer cells was associated with the increased level of TRDP. Treatment of MDA-MB-435 human breast carcinoma cells with the short telomere-specific oligonucleotide $(TTAGGG)_3$, which was shown to be an effective approach for experimental telomere elongation in immortal eukaryotic cells, significantly inhibited serum deprivation-induced apoptosis. Taken together, our data strongly suggest that telomere degradation and excessive accumulation of TRDP are closely related to apoptosis and may be directly involved in execution of programmed cell death. Our findings also imply that telomeric DNA repeats are direct targets for apoptosis-associated nuclear endonucleases, and enhanced degradation of telomeric DNA and increased levels of TRDP may serve as an intracellular signal triggering the apoptosis through amplification of chromosome instability and acceleration of apoptotic degradation of DNA. We suggest that degradation of telomeres is a common intracellular biochemical pathway of replicative senescence and apoptosis, and that programmed cell death could be defined as a catastrophic cell senescence.

4.2. Introduction

The ends of linear chromosomes of eukaryotes are composed of protein-DNA structures called telomeres.[1-4] In vertebrates, telomeres consist of hundreds to thousands of tandem repeats of the sequence (TTAGGG)*n* and associated proteins, and appear to function in chromosome protection from DNA degradation, rearrangements, and end-to-end fusions.[1-9] Telomeres are essential for chromosome integrity, positioning, complete replication, and prevention of chromosome loss. Telomeric repeats are substrates for novel replication mechanisms and usually replicated by telomerase, a telomere-specific reverse transcriptase that directs the synthesis of telomeric repeats at chromosome ends.[10-12] The analysis of the telomere lengths in normal somatic cells has shown that chromosomes lose about 50 to 200 nucleotides of telomeric sequences per cell division.[13-15] This has been attributed to the inability of conventional DNA polymerase to replicate the ends of linear DNA.[16,17] Accessibility of telomeric DNA to nucleases and nucleolytic degradation of telomeres may also influence the maintenance of telomere length.[4] For instance, in normal human somatic cells, which express low or undetectable levels of telomerase, telomeres slowly shrink during successive cell divisions in vitro or with aging in vivo.[13-15,18-22] The telomere shortening has been proposed to be a mitotic clock by which cells count their division,[19] and a sufficient shortening or elimination of telomeres may serve as a signal for replicative senescence.[14,15,18,20,23] The germline and immortal cells, as well as cancer cells and tissues, express telomerase activity and maintain telomere length through an indefinite number of cell divisions.[21,24-28] However, telomerase expression in immortalized human cells does not lead to unlimited telomere elongation.[21,24] Moreover, the telomeres that have been detected in a range of cancer cells and tissues, including breast and prostate cancers, gastrointestinal and hematological malignancies, are significantly shortened compared to the normal somatic cells.[18,27,29-36] Furthermore, the telomeres in human cells shorten at a rate ~10 times faster than that expected from incomplete replication,[21,37] suggesting that degradation of telomeric DNA occurs during cell division.

Apoptosis (programmed cell death) is an important homeostatic mechanism that balances cell production with cell death and maintains correct cell number in the body.[38-41] Aberrant cell survival resulting from inhibition of apoptosis is expected to contribute to oncogenesis and tumor progression[39-42] since cancer cells,[39-41,43] particularly metastatic cancer cells,[44] would gain a selective growth advantage by blocking the apoptosis. One of the main biochemical features that distinguishes programmed cell death is the induction of nuclear DNA fragmentation in intranucleosomal regions (characteristic ladder pattern of DNA fragmentation).[45] However, little is known about the type of chromosomal DNA that is a precise target for fragmentation during apoptosis. It is not known, for instance, whether telomeric DNA is subject to degradation during apoptosis, as well as whether or not the telomere-dependent replicative senescence and programmed cell death are related to each other. We reasoned that the analysis of degradation of telomeric DNA could be an appropriate experimental approach to establish a link between replicative cell senescence and apoptosis. Since, despite the presence of active telomerase, cancer cells have significantly shortened telomeres (see above), one may expect a higher level of telomere degradation in neoplastic cells. Furthermore, because there is an association of aggressive metastatic phenotype with resistance to apoptosis,[44] the panel of established metastatic cell lines with different metastatic potentials and different apoptosis levels appears to be a suitable cellular model to address this problem.

Here we report that telomeres manifest a specific pattern of telomeric DNA degradation during cell division in vitro, suggesting a nuclease-mediated G-tail formation mechanism on chromosome ends. We document that enhanced degradation of telomeric DNA is associated with increased activity of nuclear Ca^{2+}-dependent endonucleases and higher levels of apoptosis in a panel of nine murine and human cancer cell lines exhibiting different metastatic potentials and different apoptosis levels. Decreased degradation of telomeric DNA, low nuclear endonuclease activity, and diminished levels of apoptosis were associated with high metastatic potential. We show that induction of apoptosis in metastatic cancer cells is associated with the accumulation of telomere fragments, and experimental elongation of telomeres can protect cells from apoptosis. We suggest that degradation of telomeres is a common intracellular biochemical pathway of replicative cell senescence and apoptosis, and hypothesize that critical shortening of telomeres may trigger apoptosis by influencing the free level of telomere-binding proteins and/or derepressing gene(s) placed near the telomeres.

4.3. Identification of the Telomeric Repeat Degradation Products (TRDP) Using a Polymerase Chain Reaction (PCR)-Based Assay of Telomere Fragmentation in Murine and Human Cancer Cells

In these experiments murine and human cancer cell lines were grown in cell culture as described[44] and at day 2 the quantitative extraction and agarose gel electrophoresis of the low molecular weight DNA were made from an equal number of cells, in each case exactly as previously reported.[44] The presence of TRDP in low molecular weight DNA extracts was determined using a PCR amplification protocol with vertebrate telomere-specific primers[6] such as $(T_2AG_3)_3$ (primer 1) and $(C_3TA_2)_3$ (primer 2). The resulting PCR-amplified TRDP were analyzed by electrophoresis in an agarose gel and visualized by ethidium bromide staining. To avoid artifacts resulting from PCR primer complementarity, the interaction between two telomere-specific primers must be prevented. Hence, the single telomere-specific primer (either primer 1 or primer 2) was used in each PCR amplification assay. The PCR product formation in TRDP amplification assay is absolutely dependent on the presence of telomere-specific primer 1 (T_2AG_3). Absence of either $(T_2AG_3)_3$ oligonucleotide or DNA extract does not lead to formation of PCR product. No PCR product was identified when Taq DNA polymerase was omitted or when telomere-specific primer 2 was used. To estimate the optimal level of DNA template in the PCR amplification reaction, we carried out serial dilution of each low molecular weight DNA extract. DNA extracts were diluted and analyzed at different concentrations to determine the maximum level of TRDP amplification that could be achieved under described PCR conditions and to avoid false negatives due to possible inhibitors of PCR in the extracts. For every experimental condition, at least two separate experiments were conducted, and the results were confirmed in at least three TRDP assays.

The PCR amplification of low molecular weight DNA extracts with $(T_2AG_3)_3$ telomere-specific primer generates a PCR product comprising multiple low molecular weight DNA bands with approximate molecular weight <1 kb (Fig. 4.1). The reaction is highly specific since no PCR product was detected using either irrelevant primer or the $(C_3TA_2)_3$ telomere-specific primer. Thus, PCR-amplification analysis revealed the presence of (CCCTAA)*n* telomere fragments with molecular weight <1kb in low molecular weight DNA extracted from proliferating in vitro murine and human cancer cells. To check the size of the DNA fragments synthesized by Taq polymerase in the telomeric repeat degradation product (TRDP) assay, alkaline agarose gel electrophoresis was used.[46] Single-stranded DNA analysis in 1.2% alkaline agarose gel electrophoresis showed that the size of PCR-amplified telomere fragments is in the range of 200-600 bases (data not shown). Multiple control experiments demonstrated that a PCR positive signal in the TRDP assay required presence of the telomere-specific primer $(T_2AG_3)_3$, but not $(C_3TA_2)_3$, and DNA template, validating the TRDP assay for specific detection of telomere fragmentation (Fig. 4.1). These data are consistent with the nuclease-mediated mechanism of single-strand telomere degradation leading to formation of single-strand DNA G tail on chromosome ends. The transient formation of single-strand G tails, comprising >30 bases of G-rich DNA strand extension over C-rich DNA strand, has been detected on *Saccharomyces* chromosomes in the late S phase (reviewed in ref. 4). Consistent with this hypothesis, the PCR products of the described telomere fragments are multiple single-strand DNA fragments, containing (C_3TA_2)n structures. Alternatively, the

Fig. 4.1. (opposite page) A. Identification of the telomeric repeat degradation products (TRDP) using a polymerase chain reaction (PCR)-based assay of telomere fragmentation in murine B16 melanoma cells.
Murine B16 melanoma cell lines were grown in cell culture as described[44] and at day 2 quantitative extraction and agarose gel electrophoresis of the low molecular weight DNA were made from an equal number of cells in each case exactly as previously reported.[44] The presence of TRDP in low molecular weight DNA extracts was determined using a PCR amplification protocol with vertebrate telomere-specific primers (Meyne et al, 1989) such as $(TTAGGG)_3$ (primer 1) and $(CCCTAA)_3$ (primer 2). The resulting PCR-amplified TRDP were analyzed by electrophoresis in an agarose gel and visualized by ethidium bromide staining. To avoid artifacts resulting from PCR primer complementarity, the interaction between two telomere-specific primers must be prevented. Hence, the single telomere-specific primer (either primer 1 or primer 2) was used in each PCR amplification assay. The PCR product formation in TRDP amplification assay is absolutely dependent on the presence of telomere-specific primer 1$(TTAGGG)_3$. Absence of either $(TTAGGG)_3$ oligonucleotide or DNA extract does not lead to formation of PCR product. No PCR product was identified when Taq DNA polymerase was omitted or when telomere-specific primer 2 was used.
Lane 1, MW DNA markers (Lambda DNA/Hind III; Promega Corp., Madison, WI); Lane 2, TRDP assay with $(TTAGGG)_3$ primer of low molecular weight DNA extract from B16-F10 melanoma cells after 2 days of culture in 1% FCS; Lane 3, TRDP assay with

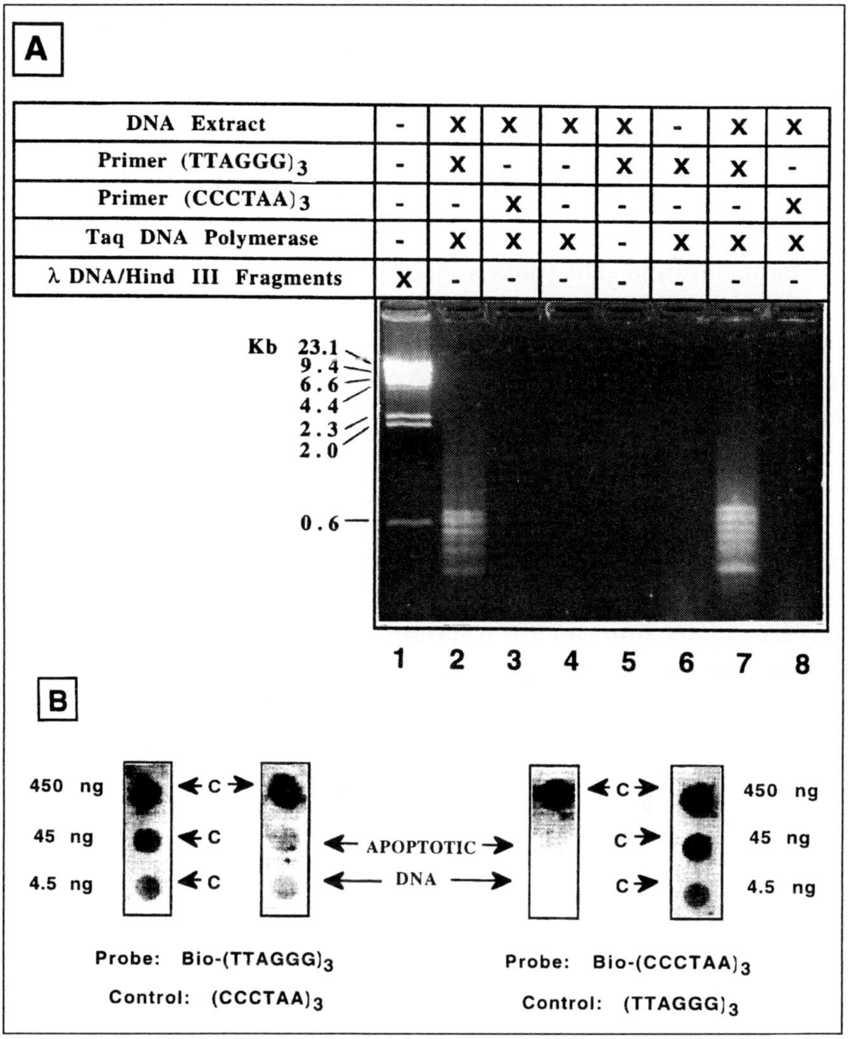

Fig. 4.1. Cont.
$(CCCTAA)_3$ primer of low molecular weight DNA extract from B16-F10 melanoma cells after 2 days of culture in 1% FCS; Lane 4, control omitting telomere-specific primers; Lane 5, control omitting Taq DNA polymerase; Lane 6, control omitting DNA extract; Lane 7, TRDP assay with $(TTAGGG)_3$ primer of low molecular weight DNA extract from B16-F1 melanoma cells after 2 days of culture in 1% FCS; Lane 8, TRDP assay with $(CCCTAA)_3$ primer of low molecular weight DNA extract from B16-F1 melanoma cells after 2 days of culture in 1% FCS. The results shown are representative of at least three separate experiments.
B. Hybridization assay of TRDP in apoptotic DNA from MDA-MB-468 human breast carcinoma cells using as a probe biotinylated telomere-specific oligonucleotides Bio-$(TTAGGG)_3$ or Bio-$(CCCTAA)_3$. C-positive controls: samples are the cold (nonlabeled) telomere-specific oligonucleotides complementary to the corresponding biotinylated probes. 40 μg of low molecular weight apoptotic DNA and indicated amounts of synthetic oligonucleotides in the control samples were loaded. See text for details.

preferential accumulation during cell division in vitro of telomere-specific oligonucleotides derived from the C-rich DNA strand may indicate the effective reutilization of telomere-specific oligonucleotides derived from the G-rich DNA strand, or their differential sensitivity to exonucleolytic degradation. However, both synthetic telomere-specific oligonucleotides showed similar sensitivity in vitro to nuclease digestion when incubated with cell extract from B16-F1 murine melanoma cells (data not shown).

4.4. Enhanced Degradation of Telomeric DNA is Correlated with the Increased Apoptosis Levels in Metastatic Cancer Cells

We have established earlier that growth in vitro to high-saturation density as well as following serum deprivation induces apoptosis in metastatic cancer cells.[44,78,79] By using PCR amplification with telomere-specific primers, metastatic cancer cells were examined for the presence of telomeric DNA fragments in low molecular weight DNA from Hirt supernatant fractions isolated before and after induction of apoptosis. The procedure permitted the quantitative recovery of low molecular weight DNA from high molecular weight chromosomal DNA from metastatic cells at different time points before and during maximal apoptosis.

PCR-based TRDP assay showed a clear association of the level of accumulation of telomere fragments with the level of apoptosis in metastatic cancer cells (Fig. 4.2). Highly metastatic cancer cells, which exhibited a significantly lower level of apoptosis compared to their poorly metastatic counterparts,[44,78] manifested also a substantially lower level of telomere fragmentation (Fig. 4.2C,D). Induction of apoptosis in metastatic cancer cells by serum deprivation was accompanied by a substantial increase in the levels of accumulation of telomere fragments (Fig. 4.2A,B). The relative levels of low molecular weight telomere fragments in metastatic cancer cells at 24 h and 48 h after plating were a good indicator of subsequent cell survival and apoptosis levels in cultures growing to high saturation density or following serum deprivation (Fig. 4.2B-D). Thus, the levels of degradation of telomeric DNA in murine and human cancer cells exhibited a direct correlation with the apoptosis levels during growth to high saturation density or serum deprivation.

4.5. Enhanced Degradation of Telomeric DNA is Associated with Activation of Nuclear Ca^{2+}-Dependent Endonucleases

Activation of nuclear Ca^{2+}-dependent endonucleases is responsible for the DNA fragmentation during apoptosis.[45,60,61] We therefore investigated whether Ca^{2+}-dependent endonuclease activity in isolated nuclei from human melanoma and breast carcinoma cells is correlated with the apoptosis level, DNA fragmentation and degradation of telomeres. High levels of apoptosis, DNA fragmentation, and telomere degradation in poorly metastatic DX-3 melanoma and MDA-MB-468 breast carcinoma cells were associated with high levels of endogenous endonucleases, available for direct activation by Ca^{2+} in isolated nuclei (Fig. 4.3). In contrast, highly metastatic MDA-MB-435 breast carcinoma and TXM-13 melanoma cells that exhibit low levels of apoptosis, DNA fragmentation, and telomere degradation also express little to none of nuclear Ca^{2+}-dependent endonuclease activity (Fig. 4.3). Thus, the levels of apoptosis, DNA fragmentation, and telomere degradation in metastatic cancer cell lines are correlated with the activity of nuclear Ca^{2+}-dependent endonucleases suggesting that telomeric DNA could be a direct target for nuclear endonucleases involved in execution of programmed cell death. To test this hypothesis we performed hybridization of low molecular weight apoptotic DNA extracted from either cells or isolated nuclei with biotinylated telomere-specific oligonucleotides (Fig. 4.4). As expected, significantly stronger TTAGGG telomere-specific hybridization signals were detected in low molecular weight DNA extracts from either cells subjected to serum-deprivation induced apoptosis or calcium-treated nuclei (Fig. 4.4). In agreement with the PCR-based TRDP assay data, no hybridization signals were detected with biotinylated CCCTAA telomere-specific oligonucleotides (Fig. 4.4). Thus, these results strongly support the idea that telomeric DNA repeats are direct targets for nuclear Ca^{2+}-dependent endonucleases activated during apoptosis.

4.6. Caspase Activation is not Necessary for Degradation of Telomeric DNA

Experimental evidence has suggested a high degree of conservation of the ICE-like protease (caspase)-dependent pathway of apoptosis among species and various initiators of programmed cell death, including TNF-α and Fas ligand.[59,72-76] We next examined the effect of specific ICE-like protease inhibitor peptide N-benzyloxy-

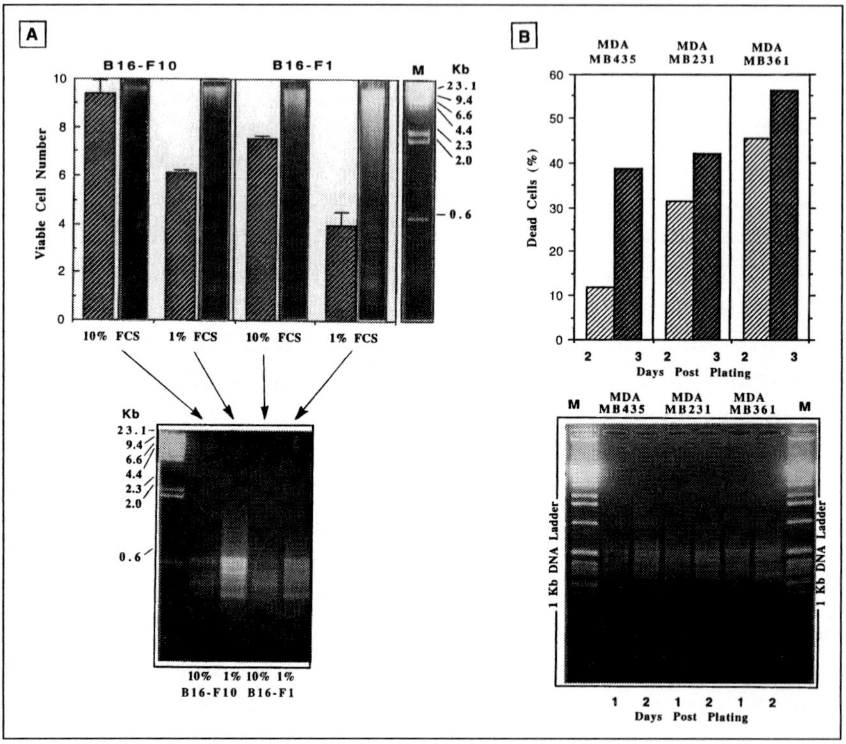

Fig. 4.2. Apoptosis levels in human and murine metastatic cancer cell lines are correlated with the accumulation of telomere fragments.
A & B. Induction of apoptosis in human and murine metastatic cancer cells by serum deprivation is accompanied by the accumulation of telomere fragments. (A) Serum deprivation-induced apoptosis in murine B16 melanoma cell lines is associated with decreased cell viability, accumulation of apoptotic DNA (top panel), and increased level of telomere degradation (bottom panel). (B) Correspondence between the levels of telomere fragmentation (bottom panel) and subsequent cell death (top panel) of human breast carcinoma cell lines subjected to serum deprivation-induced apoptosis.
C & D. Apoptosis levels induced by growth to saturation density in human and murine metastatic cancer cell lines are correlated with the accumulation of telomere fragments. (C) Apoptosis level as determined by the TUNEL assay (top and middle panels) or DNA fragmentation analysis (bottom left panel) and accumulation of telomere fragments (bottom right panel) as determined in TRDP assay in highly (B16-BL6) and poorly (B16-F1) metastatic murine melanoma cell variants. (D) Apoptosis level as determined by the TUNEL assay (top panel) and cell viability assayed by trypan blue exclusion (middle panel) and accumulation of telomere fragments (bottom panel) as determined in TRDP assay in a panel of human breast carcinoma cell lines exhibiting high (MDA-MB-435), intermediate (MDA-MB-231), and low (MDA-MB-361) metastatic potential in nude mice (see text).
Cell viability in medium containing either 10%, 1% or 0.1% FCS was assayed by trypan blue exclusion as previously described.[44] Identification of apoptotic cells employing the TUNEL method (terminal deoxynucleotidyl transferase-mediated dUTP-X nick end labeling method) was performed exactly as reported earlier.[44] The In Situ Cell Death Detection Kit, POD (Boehringer Mannheim Co., Indianapolis, IN) was employed, using fluorescein-dUTP and an anti-fluorescein antibody conjugated to a peroxidase

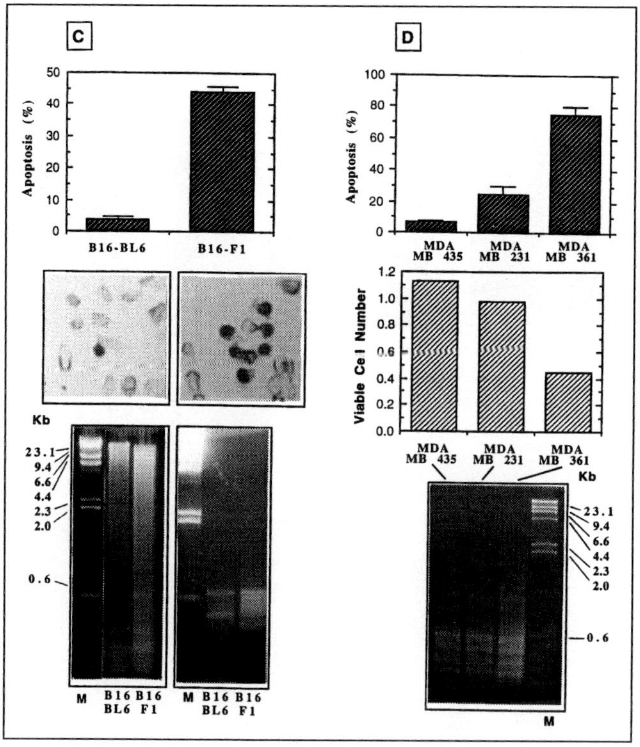

Fig. 4.2 Cont.
reporter molecule as detection reagent. DNA fragmentation analysis was performed at day 2 in each culture exactly as described.[44] The accumulation of telomere fragments was determined in the TRDP assay as outlined in the legend to Figure 4.1. Cell culture conditions in both serum deprivation (0.1% FCS for human breast carcinoma cell lines and 1% FCS for murine melanoma cell variants) and saturation density (medium containing 10% FCS) experiments were similar to those previously reported.[44] The panels of murine melanoma cell variants and human breast carcinoma cell lines exhibiting different metastatic potential and different sensitivity to apoptosis were previously described (see text). Percent of apoptotic cells, actual average viable cell numbers (x10^6) or percent viability are plotted. For every experimental condition, at least two separate experiments were conducted, and the results were confirmed in at least two TRDP assays.

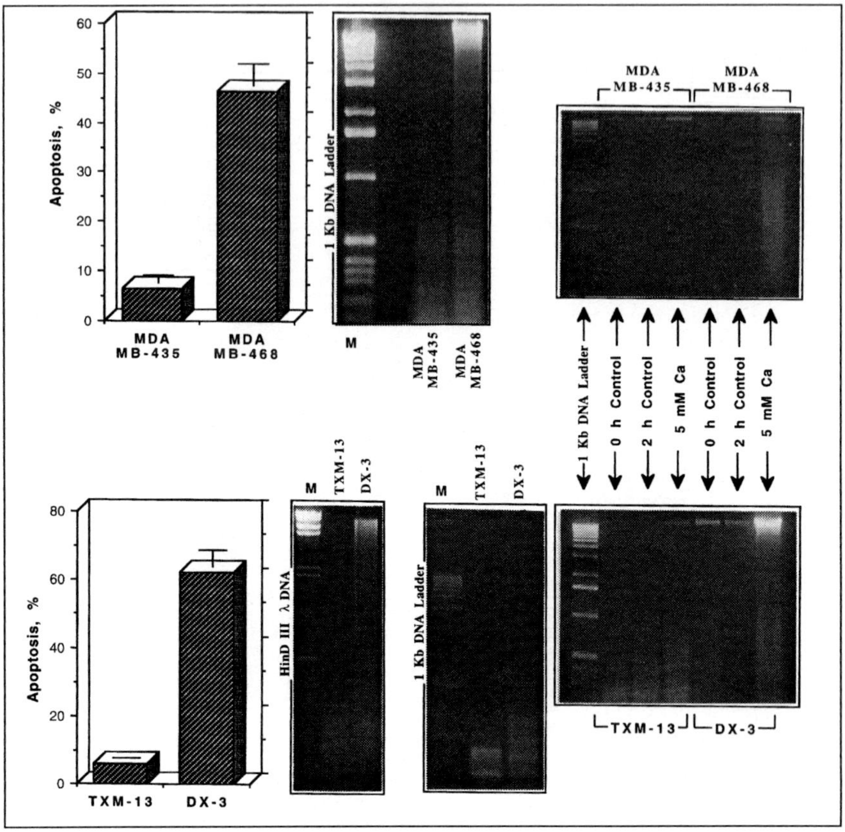

Fig. 4.3. Correlation of the levels of apoptosis, DNA fragmentation, and telomere degradation in human breast carcinoma (top panel) and melanoma (bottom panel) cells with the availability of endogenous nuclear Ca^{2+}-dependent endonucleases for direct in vitro activation by calcium in isolated nuclei.
Cells were harvested from subconfluent cultures (60-70% confluence), plated at $1x10^5$ ml^{-1} in multiple flasks and cultivated for 6 days in medium containing 10% FCS. Cell viability was assayed by acridine orange staining and/or trypan blue exclusion as described above (see legend to Fig. 4.2). Identification of apoptotic cells employing the TUNEL method (terminal deoxynucleotidyl transferase-mediated dUTP-X nick end labeling method) was performed exactly as reported earlier.[44] DNA fragmentation analysis (top panel, middle figure and bottom panel, middle left figure) was performed at day 2 (breast carcinoma) and 6 (melanoma) in each culture exactly as described (see legend to Fig. 4.2). The accumulation of telomere fragments (bottom panel, middle right figure) was determined in the TRDP assay as outlined in the legend to Figure 4.1. Cell culture conditions were identical to those previously reported.[44,78] Percent of apoptotic cells is plotted. For every experimental condition, at least two separate experiments were conducted, and the results were confirmed in at least two TRDP assays.
Nuclei were isolated from subconfluent cultures (60-70% confluence) and incubated for 2 hours in the absence or presence of 5 mM Ca^{2+}. DNA fragmentation in low molecular weight DNA extracts from identical numbers of control and calcium-treated nuclei (top and bottom panels, right figures) was determined by agarose gel electrophoresis as previously described.[44]

carbonyl-Val-Ala-Asp-fluoromethylketone (Z-VAD-fmk) on serum deprivation-induced apoptosis and telomere degradation. Treatment of the MDA-MB-435 human breast carcinoma cells with 50 μM concentration of the ICE-like protease inhibitor Z-VAD-fmk almost completely blocked the serum deprivation-induced apoptosis (Fig. 4.5). Thus, functional ICE-like protease (or proteases) is required for execution of serum deprivation-induced apoptosis in metastatic human breast carcinoma cells. Among the broad family of the apoptosis-associated ICE-like proteases in mammalian cells, the CPP32/Yama death protease is considered as one of the key enzymes largely responsible for degradation of several critical nuclear protein targets during apoptosis.[72-75] The cleavage of the nuclear protein death substrates such as poly (ADP-ribose) polymerase (PARP) and the 70 kDa protein component of the U1 small nuclear ribonucleoprotein (U1-70 kDa) by the CPP32/Yama may play a central role in apoptosis execution since it will terminate DNA repair and prevent mRNA splicing.[74-76] To directly confirm the ICE-like protease activation during apoptosis in MDA-MB-435 cells, we next assessed the activity of the ICE and CPP32/Yama death proteases employing corresponding specific peptide substrate for ICE and CPP32/Yama, synthetic peptides Ac-YVAD-pNA and Ac-DEVD-pNA (Calbiochem). To eliminate the potential interference of nonspecific hydrolysis due to contaminating proteinases,[77] an assay was performed using 100 μM of the specific peptide inhibitor an of ICE and CPP32/Yama, peptides Ac-YVAD-CMK and Ac-DEVD-CHO, respectively. As shown in Fig. 4.6B, the induction of apoptosis by serum deprivation in MDA-MB-435 cells was accompanied by more than three-fold elevation in the activity of CPP32/Yama at 48 hours. In contrast, no significant increase of the ICE activity was detected (Fig. 4.6B). Thus, induction of apoptosis in MDA-MB-435 human breast carcinoma cells was associated with significant elevation of the activity of the CPP32/Yama death protease. Activation of the CPP32/Yama death protease may play an important role in execution of apoptosis in MDA-MB-435 cells, since treatment with ICE-like protease inhibitor peptide Z-VAD-fmk prevent apoptosis induction.

It is well known that specific telomere-binding proteins protect telomeres from nucleolytic degradation (reviewed in ref. 4). Thus, telomere fragmentation during apoptosis could be accelerated as a result of ICE-like death protease activation and cleavage of telomere-binding proteins. We next assessed the effect of ICE-like

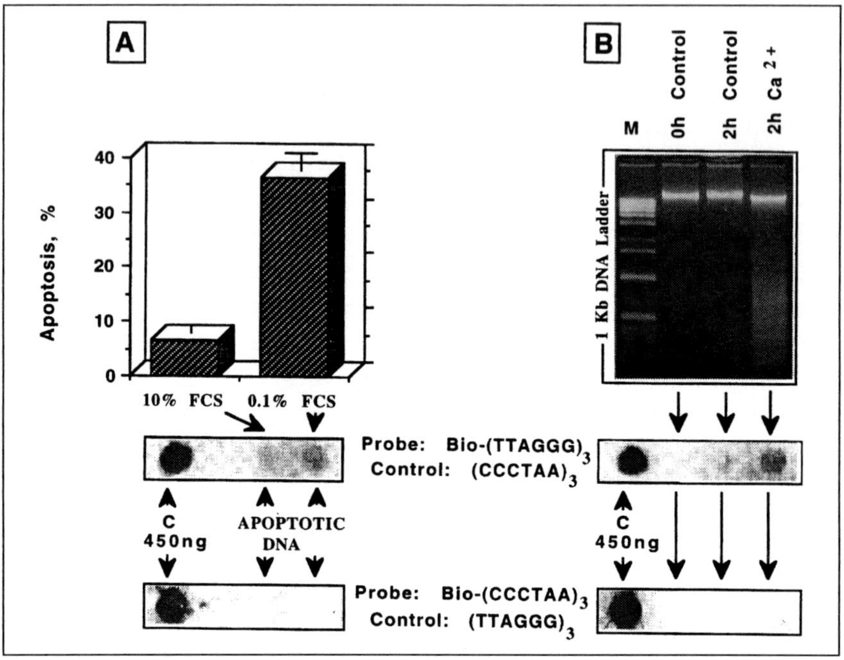

Fig. 4.4. Hybridization analysis of the accumulation of TRDP in low molecular weight apoptotic DNA during apoptosis induced either by serum deprivation in the cells (A) or by 5 mM calcium in isolated nuclei (B).
A. MDA-MB-435 human breast carcinoma cells were cultivated in medium containing either 10% or 0.1% FCS and assayed for apoptosis employing the TUNEL methods. At day 2 quantitative extraction of the low molecular weight DNA was made and identification of the telomere fragmentation was performed as described in the legend to Figure 4.1.
B. Nuclei were isolated from B16-F1 murine melanoma cells grown to 60%-70% confluence in 10% FCS. 3 x 10^6 nuclei were incubated at 37°C for 2 h in TBS with and without addition of 5 mM Ca^{2+}; low molecular weight DNA was quantitatively extracted and assayed for TRDP employing hybridization protocol as described in the legend to Figure 4.1.
Hybridization assay of TRDP was made using as a probe biotinylated telomere-specific oligonucleotides Bio-$(TTAGGG)_3$ or Bio-$(CCCTAA)_3$. C-positive controls; samples are the cold telomere-specific oligonucleotides complementary to the corresponding biotinylated probes.

protease inhibitor Z-VAD-fmk peptide on telomere degradation and found that Z-VAD-fmk-treated cells manifested essentially the same level of telomere fragmentation as determined in PCR-based TRDP assay (Fig. 4.5). These results indicate that associated with serum deprivation-induced apoptosis increase in telomere fragmentation and accumulation of TRDP does not require a functional ICE-like protease and rather depends upon direct nuclease activation at the

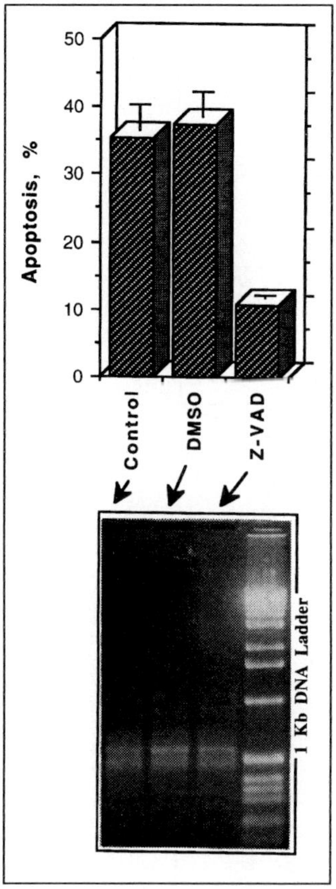

Fig. 4.5. Effect of specific ICE-like protease inhibitor peptide N-benzyloxycarbonyl-Val-Ala-Asp-fluoromethylketone (Z-VAD-fmk) on serum deprivation-induced apoptosis and telomere degradation in MDA-MB-435 cells.
Cells were harvested from subconfluent cultures (60-70% confluence), plated at $4x10^5 ml^{-1}$ in multiple flasks with and without specific peptide inhibitor of ICE-like protease Z-VAD-fmk (50 μM; Enzyme System Products, Dublin, CA), and cultivated for 3 days in medium containing 0.1% FCS. Cell viability was assayed by acridine orange staining and/or trypan blue exclusion as described above (see the legend to Fig. 4.2). Identification of apoptotic cells employing the TUNEL method (terminal deoxynucleotidyl transferase-mediated dUTP-X nick end labeling method) was performed exactly as reported earlier.[44] The accumulation of telomere fragments (bottom panel) was determined in the TRDP assay as outlined in the legend to Figure 4.1. Cell culture conditions in serum deprivation experiments (0.1% FCS for human breast carcinoma cells) were identical to those previously reported.[44,78] Percent of apoptotic cells, actual average viable cell numbers ($x10^{-6}$) or percent viability are plotted. For every experimental condition, at least two separate experiments were conducted, and the results were confirmed in at least two TRDP assays.

early stage of apoptosis induction. These experiments also imply that increased telomere degradation is not sufficient for apoptosis execution when ICE-like protease activity is suppressed.

4.7. CPP32/Yama Death Protease Activation does not Lead to Apoptosis In the Cells Treated with Short Telomere-Specific Oligonucleotides TTAGGG

It has been established that treatment of immortal human cell lines in vitro with short telomere-specific oligonucleotides TTAGGG resulted in telomere elongation.[49] To test the hypothesis that elongation of telomeres can protect cells from apoptosis, the MDA-MB-435 human breast carcinoma cells were treated with short telomere-specific oligonucleotides and subjected to induction of apoptosis in 0.1% FCS (Fig. 4.6A). Cells treated with $(TTAGGG)_3$ oligonucleotides manifested a marked reduction in apoptosis compared to untreated control cells. The effect is highly specific since no reduction of

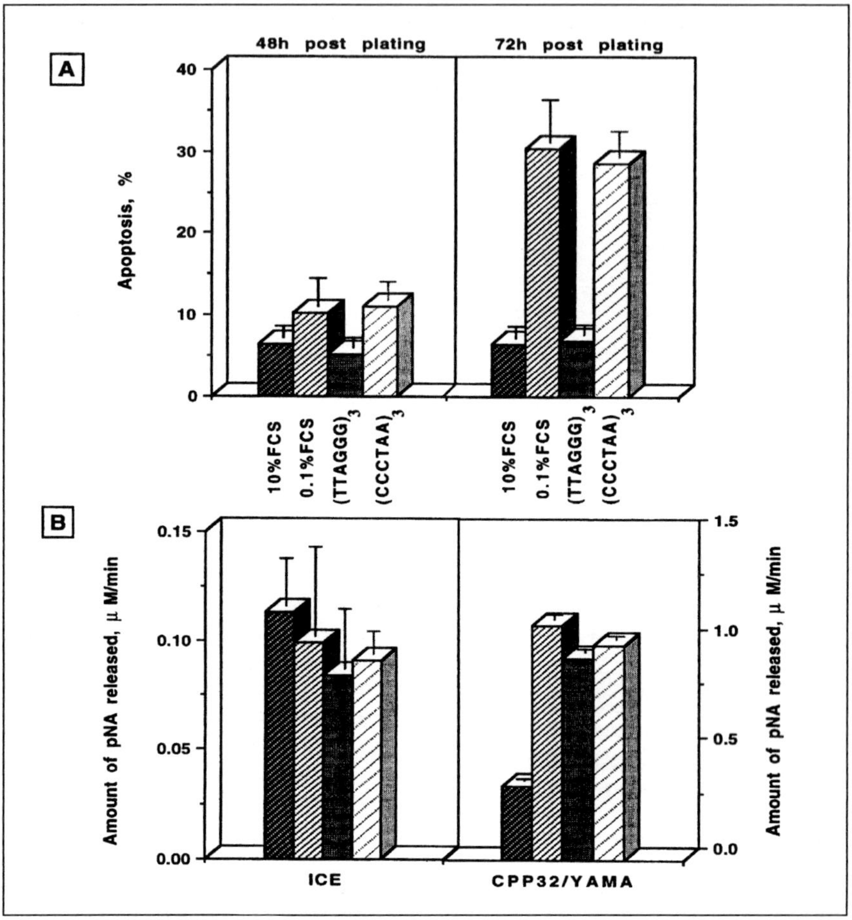

Fig. 4.6. Effect of telomere-specific oligonucleotides $(TTAGGG)_3$ and $(CCCTAA)_3$ on serum deprivation-induced apoptosis in MDA-MB-435 human breast carcinoma cells. A. Inhibition of apoptosis by $(TTAGGG)_3$ telomere-specific oligonucleotides. Cells were harvested from subconfluent cultures (60-70% confluence), plated at $2x10^5$ ml^{-1} in multiple flasks with and without synthetic telomere-specific oligonucleotides (4.5 µg/ml final concentration), and cultivated for 3 days in medium containing 0.1% FCS. Cell viability was assayed by acridine orange staining and/or trypan blue exclusion as described above (see the legend to Fig. 4.2). Identification of apoptotic cells employing the TUNEL method (terminal deoxynucleotidyl transferase-mediated dUTP-X nick end labeling method) was performed exactly as reported earlier.[44] Cell culture conditions in serum deprivation experiments (0.1% FCS for human breast carcinoma cells) were similar to those previously reported.[44] Percent of apoptotic cells is plotted. For every experimental condition, at least two separate experiments were conducted.
B. Effect of telomere-specific oligonucleotides on ICE and CPP32/Yama death protease activities during serum deprivation-induced apoptosis.
Cells were cultivated for 2 days as described above (see the legend to Fig. 4.6A), harvested and assayed for ICE and CPP32/Yama activities as described in ref. 78. The representative results of one of two separate experiments are shown.

apoptosis level was observed in cells treated with complementary telomere-specific $(CCCTAA)_3$ oligonucleotides (Fig. 4.6A). It should be noted that a single application of the $(TTAGGG)_3$ telomere-specific oligonucleotides at $\sim 10^{-6}$ M was sufficient to cause a significant increase in viability of MDA-MB-435 cells and over 80% inhibition of serum deprivation-induced apoptosis at 72 hours after treatment.

Since activation of CPP32/Yama death protease is required for serum deprivation-induced apoptosis in MDA-MB-435 cells (see above), we next examined whether inhibition of apoptosis by telomere-specific TTAGGG oligonucleotides would interfere with CPP32/Yama activation. We determined that TTAGGG-treated MDA-MB-435 cells growing in the medium containing 0.1% FCS display the same degree of the CPP32/Yama death protease activation as untreated cells subjected to serum deprivation-induced apoptosis (Fig. 4.6B). Thus, inhibition of apoptosis by telomere-specific TTAGGG oligonucleotides does not interfere with serum deprivation-induced CPP32/Yama activation and, therefore, does not require blocking of the ICE-like protease-dependent pathway of apoptosis execution. These data also imply that CPP32/Yama death protease activation is necessary but not sufficient for execution of serum deprivation induced apoptosis in MDA-MB-435 human breast carcinoma cells treated with telomere-specific TTAGGG oligonucleotides.

4.8. Cells with Experimentally Elongated Telomeres Manifest Higher Resistance Toward Apoptosis

A single application of telomere-specific TTAGGG oligonucleotides, which causes a marked inhibition of apoptosis, does not induce a significant detectable experimentally telomere elongation (ref. 49). To independently test the hypothesis that telomere length may influence sensitivity to apoptosis, we compared the response to serum deprivation-induced apoptosis of the established prostate adenocarcinoma cell line GTR-DU145 and immortal human fibroblast cell line GTR-IDH4 with stable experimentally elongated telomeres and parental cell lines DU145 and IDH4 (ref. 49). The levels of serum deprivation-induced apoptosis as determined by the TUNEL assay was two-fold higher in the DU145 cells compared to GTR-DU145 cells (Fig. 4.7A) and four-fold higher in IDH4 cells compared to GTR-IDH4 cells (ref. 80). We also found that GTR-DU145 and GTR-IDH4 cells exhibit a significantly higher viability compared to DU145 and IDH-4 cells when cell lines were growing in 0.1% FCS (data not shown). Thus, the GTR-DU145 and GTR-IDH4 cells that are phenotypically different from parental DU145 and IDH4 cells only in telomere length manifest a significantly higher resistance toward apoptosis.[80] Taken together, these results provide strong direct experimental evidence for the hypothesis that telomere status may control sensitivity to apoptosis and suggest that the level of low molecular weight telomere fragments may directly influence the execution of programmed cell death.

4.9. Two Major Apoptosis-Execution Mechanisms are Deficient in Cells with Experimentally Elongated Telomeres

The execution of apoptosis requires an activation of two main degradation pathways that are largely responsible for destruction of key nucleic acid and protein death substrates: Ca^{2+}-dependent nuclear endonucleases and ICE-like proteases (caspases). We next determined whether higher resistance toward apoptosis of GTR-DU145 and GTR-IDH4 cells is associated with aberrations in apoptosis execution mechanisms (ref. 80). The levels of low molecular weight apoptotic DNA were substantially lower in the GTR-DU145 and GTR-IDH4 cells compared to the parental DU-145 and IDH4 cells in serum deprivation experiments (Fig. 4.7D), suggesting a deficiency of nuclear Ca^{2+}-dependent endonucleases in GTR-DU145 cells. Indeed, the availability of nuclear endonucleases for direct activation by calcium was diminished in nuclei isolated from GTR-DU145 cells (Fig. 4.7E). Furthermore, the levels of the ICE and CPP32/Yama death proteases were significantly lower in GTR-DU145 compared to DU-145 prostate carcinoma cells (Fig. 4.7B). Growth to high density of DU-145 cells was accompanied by substantial increase in the activity of the ICE death protease (Fig. 4.7C). However, in contrast to the DU-145 cells this increase in the ICE activity was not seen in the GTR-DU145 cells in saturation density experiments. Similarly, growth in 0.1% FCS of IDH4 cells was accompanied by substantial increase in activity of the CPP32/Yama death protease. In contrast, this increase in CPP32/Yama activity was not seen in identically treated GTR-IDH4 cells (ref. 80). Thus, the activities of both nuclear Ca^{2+}-dependent endonucleases as well as ICE-like death proteases (caspases) are significantly diminished in the GTR-DU145 cells with stably experimentally elongated telomeres. This deficiency of two major apoptosis-execution mechanisms may contribute to the

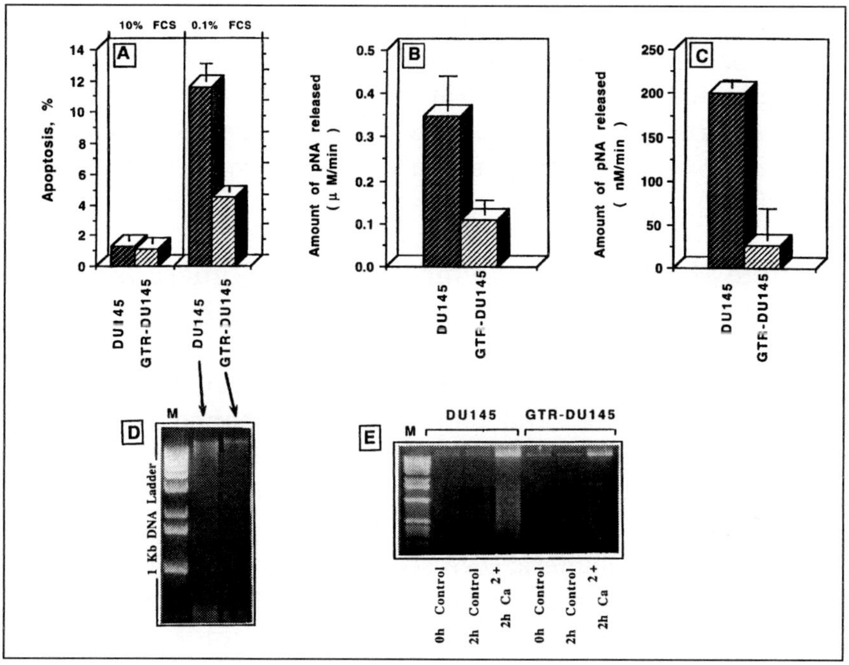

Fig. 4.7. Increased resistance toward apoptosis of cells with experimentally elongated telomeres.
Parental DU145 human prostate carcinoma cells and GTR-DU145 cells with experimentally elongated telomeres[49] were harvested from subconfluent cultures (60-70% confluence), plated at $2x10^5$ ml^{-1} (0.1% FCS) or $1x10^5$ ml^{-1} (10% FCS) in multiple flasks and cultivated for up to 7 days. Cell viability was assayed by acridine orange staining and/or trypan blue exclusion as described above (see the legend to Fig. 4.2). Cell culture conditions were similar to those previously reported.[44,78] For every experimental condition, at least two separate experiments were conducted.
A. Identification of apoptotic cells was performed at day 2 (0.1% FCS) and 6 (10% FCS) employing the TUNEL method (terminal deoxynucleotidyl transferase-mediated dUTP-X nick end labeling method) performed exactly as reported earlier.[44] Percent of apoptotic cells is plotted.
B. The ICE activity levels in DU145 and GTR-DU145 cells at day 2 post plating in 10% FCS.
C. The increase in ICE activity in DU145 and GTR-DU145 cells during growth to high saturation density. Cells were harvested at day 2 and 6 (10% FCS) and assayed for ICE death protease activity as described.[78] The results are presented as the difference in ICE activity determined at day 6 and 2 for the corresponding cell line.
D. DNA fragmentation analysis in DU145 and GTR-DU145 cells subjected to serum deprivation-induced apoptosis. Cells were harvested at day 2 (0.1% FCS) and low molecular weight apoptotic DNA was quantitatively extracted, analyzed on 1.2% agarose gels and visualized by ethidium bromide staining.
E. Activation of nuclear endonucleases by calcium is deficient in GTR-DU145 cells compared to the parental DU145 cells. Nuclei were isolated from subconfluent cultures (60-70% confluence) and incubated for 2 hours in the absence or presence of 5 mM Ca^{2+}. DNA fragmentation in low molecular weight DNA extracts from identical numbers of control and calcium-treated nuclei was determined by agarose gel electrophoresis with ethidium bromide staining.

higher resistance toward apoptosis of cells with elongated telomeres. These data also imply that telomere status, particularly telomere length, may control the sensitivity to apoptosis by influencing the activities of apoptosis-associated endonucleases as well as death proteases, possibly through mechanisms related to the telomere position effect.

4.10. Role of Telomerase, Telomeres, and Telomere-Binding Proteins in Regulation of Apoptosis Execution: A Hypothesis

Two groups of independent experimental evidence strongly imply that telomeres and telomere maintenance machinery (telomerase, telomere-binding proteins, and telomere-specific endonucleases) are directly involved in regulation of sensitivity to apoptosis and apoptosis execution. These two groups of evidence indicated that: (1) telomeric DNA sequences and components of telomere maintenance machinery are key nucleic acid and protein death substrates; (2) sensitivity toward apoptosis could be dramatically influenced by telomere-associated differences in cellular phenotypes caused by a variety of factors such as telomerase activation, telomere elongation, and treatment of cells with telomere-specific oligonucleotides. Most importantly, these two groups of facts were derived from five independent experimental approaches employing different cellular systems.

1. Induction of apoptosis in murine B16-F1 melanoma cells was accompanied by dramatic decline in telomerase activity. Similar decline in telomerase activity during apoptosis was observed in DU-145 human prostate carcinoma cells.[80]
2. Induction of apoptosis in several murine and human cancer cell lines was accompanied by significant increase in degradation of telomeric DNA. Degradation of telomeres and accumulation of TRDP was associated with activation of nuclear Ca^{2+}-dependent endonucleases.
3. The immortal telomerase-positive cell clone SW39I exhibits higher resistance toward serum starvation-induced apoptosis compared to immortal telomerase-negative cell clones SW13 and SW26IF derived from the same parental cells.[80]
4. A single application of short telomere-specific oligonucleotides TTAGGG significantly inhibits serum starvation-induced apoptosis in MDA-MB-435 human breast carcinoma cells.

5. GTR-DU145 and GTR-IDH4 cells with experimentally elongated telomeres manifest substantially higher resistance toward apoptosis and significantly reduced activity of two major apoptosis-execution mechanisms, death protease (caspase) activation and induction of nuclear Ca^{2+}-dependent endonucleases, compared to the otherwise identical parental cells, human prostate adenocarcinoma cell line DU145 and human immortal fibroblast cell line IDH4.[80] Our current understanding of the telomere-associated pathways of regulation of sensitivity toward apoptosis and control of apoptosis execution is summarized in Figure 4.8.

4.11. Conclusion: The Possible Molecular Mechanisms and Consequences of Telomere-Dependent Control of Apoptosis Execution

The precise structure of telomeric DNA at the very ends of the chromosomes is unknown. However, it has been suggested that, similarly to some protozoans, a short 3' overhang of the G-rich strand is a conservative feature of eukaryotic chromosome ends.[4,64] Due to the polarity of conventional DNA polymerase, the current model predicts that only half of the telomeres will have a gap on the 5' end of the newly synthesized C strand, thus exposing a postulated short G-rich tail. The other half of the telomeres are expected to be blunt ended.[4,64,67] Since both telomerase and telomere-binding proteins require a short overhang of the G-rich strand for binding to telomeres, the blunt end must be processed to generate this structure.[4,64-67] Recent data strongly suggest that this blunt end-processing mechanism is an integral step in telomere maintenance and involves a cell cycle-regulated selective degradation of the C-rich telomere strand by a strand-specific nuclease.[67] Our data provide substantial experimental evidence supporting this hypothesis and indicating that mammalian cells also employ such a mechanism.

The PCR-based TRDP assay generates the product comprising multiple single-strand DNA fragments in the range of 200-600 bases, containing (TTAGGG)*n* structures. Since PCR products are complementary to the telomere fragments that are present in low molecular weight DNA extracts, these findings indicate that telomere degradation products consist of multiple low molecular weight single-strand DNA fragments, containing (CCCTAA)*n* structures. These data are in agreement with the nuclease-mediated mechanism of single-strand telomere degradation leading to formation of single-

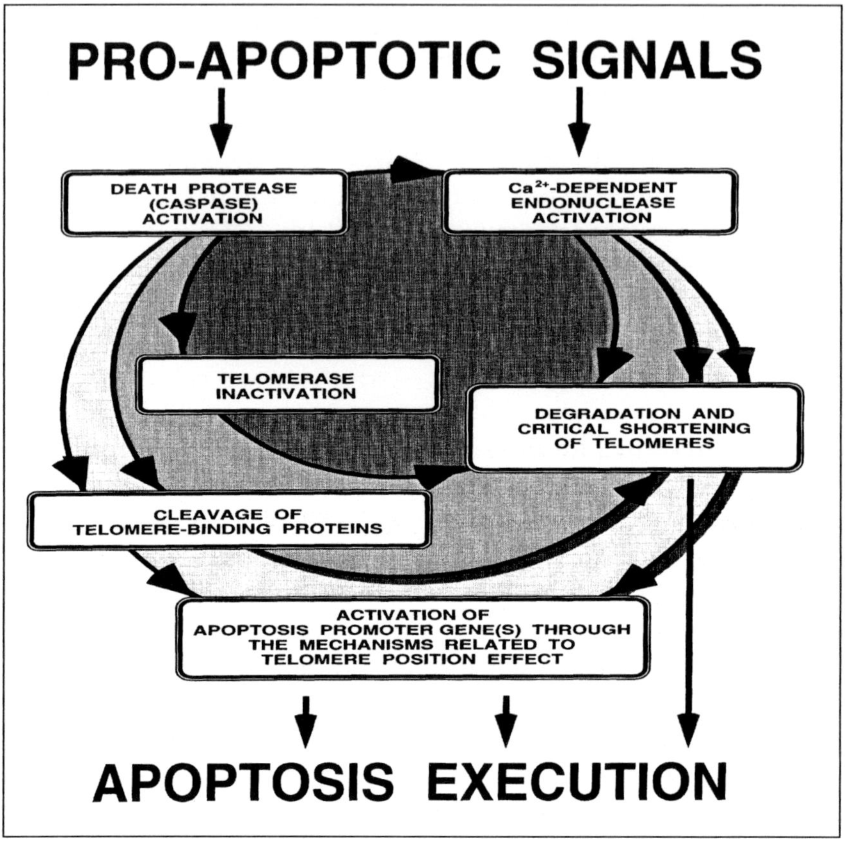

Fig. 4.8. Telomere-associated pathways of regulation of sensitivity toward apoptosis and control of apoptosis execution. See explanation in the text.

strand DNA G-rich tail on chromosome ends. The transient formation of single-strand G tails, comprising >30 bases G-rich DNA strand extension over C-rich DNA strand, has been detected on *Saccharomyces* chromosomes in the late S phase.[4,65,66] Consistent with this hypothesis, no PCR products were detected in multiple control experiments employing $(CCCTAA)_3$ telomere-specific primer. Thus, the TRDP assay revealed an asymmetrical accumulation of $(CCCTAA)n$ telomere fragments in proliferating in vitro human and murine melanoma and breast carcinoma cells. One of the intriguing features of the C-rich strand specific telomere sequence directed nuclease(s) is its apparent endonuclease activity. This may indicate that not only telomere ends are targets for degradation, but intrachromosomal telomeric DNA sequences would also be cleaved by this enzyme. It would be of interest to determine whether any similarity or functional overlap exists between telomere-specific endonuclease(s) and a structure-specific DNA endonuclease associated directly with the replication fork or DNA replication machinery.[68] On the other hand, this telomere-specific endonuclease(s) is clearly associated with apoptosis, which may indicate that domain-size apoptotic DNA fragmentation occurs at the intrachromosomal telomeric DNA tracts.

One of the obvious molecular consequences of the formation of a single-strand G-tail extension on chromosome ends would be the changes in the balance of free and DNA-bound forms of specific telomere-binding proteins that could lead to activation of telomerase and telomere elongation.[63] A similar titration effect on single-strand specific telomere-binding protein may cause an experimental telomere elongation in immortal cells treated with short telomere-specific oligonucleotides (TTAGGG)*n* (ref. 49 and this study). This mechanism is an attractive candidate for at least two important biological functions: (1) the signal for telomerase activation and telomere elongation associated with cell division; (2) the signal indicating the extent of DNA damage leading to activation of DNA repair systems or cell suicide. Consistent with this hypothesis, we observed a significant inhibition of apoptosis in MDA-MB-435 cells treated with short telomere-specific oligonucleotides $(TTAGGG)_3$. In the telomerase RNA gene from the yeast *Kluyveromyces lactis*, specific mutations that alter telomeric DNA sequences and interactions of telomeric DNA with telomere-binding proteins result in telomeres elongating to up to 100 times their normal length, with impairment of cell growth.[63] The extreme lengthening of altered telomeres was invariably accompanied by poor cell growth, characterized by rough, variably sized colonies and a high percentage of inviable cells.[63] Thus, either extreme shortening of telomeres or elongation of mutated telomeres with altered binding ability with respect to the specific telomere-binding proteins would cause an accumulation of the free form of such proteins which, in turn, may trigger an execution of the cell suicide program. In this context, it will be interesting to examine the possible functional interactions of telomere-binding proteins with p53, Rb, and Bcl-2 family members.

The data presented here establish a link between the mechanisms of telomere-dependent replicative senescence and programmed cell death. Degradation of telomeric DNA as evidenced by accumulation of telomere fragments appears to be an early event, since it could be detected before the manifestation of apoptosis as measured by the TUNEL assay, and activation of ICE-like protease(s) was not required for telomere degradation. However, the telomere degradation may act as an upstream activation signal for ICE-like protease(s) in the apoptotic cascade, since telomere-specific oligonucleotides $(TTAGGG)_3$ significantly inhibited serum deprivation-induced apoptosis, which requires an active ICE-like protease(s). However, at least for serum deprivation-induced apoptosis in MDA-MB-435 cells, this mechanism appears to be unlikely, because inhibition of apoptosis by telomere-specific TTAGGG oligonucleotides did not prevent the CPP32/Yama death protease activation. On the other hand, the protein components of telomere maintenance machinery such as telomerase and telomere-binding proteins may be the targets for ICE-like protease(s), particularly CPP32/Yama death protease. Consistent with this notion, we observed a dramatic loss of telomerase activity in B16-F1 melanoma cells subjected to serum deprivation-induced apoptosis (Glinsky GV, Holt SE, Glinsky VV, Shay JW, manuscript in preparation). Our data indicate that both telomere degradation and CPP32/Yama death protease activation are essential and necessary components of the apoptotic cascade, but neither one alone is sufficient for apoptosis execution. Therefore, the most likely molecular scenario would be that degradation of telomeres makes a specific potential death substrate(s) such as telomere-binding proteins available for cleavage by activated CPP32/Yama death protease. The cleavage of telomere-binding proteins in turn would accelerate DNA degradation and trigger the apoptosis execution. Thus, the maintenance of telomere length as a balance of their degradation and synthesis seems to be crucial in determining the fate of the cells, and apoptosis could be defined as a catastrophic cell senescence.

The critical shortening or loss of telomeres could trigger the apoptosis through two most plausible mechanisms: (1) elimination of telomere position effect[4,23] and derepression of gene(s) placed near the telomeres; (2) augmentation of the free level of telomere-binding proteins, which in turn may induce the apoptosis. Consistent with the former mechanism, we observed an increased resistance toward apoptosis and apparent deficiency of two major apoptosis execution pathways (induction of the Ca^{2+}-dependent nuclear endonucleases and activation of the ICE-like death proteases) in GTR-DU145 cells with experimentally elongated telomeres. These data may indicate that the activities of both apoptosis-associated nuclear endonucleases, as well as death proteases, are under control of a regulatory mechanism related to the telomere position effect. Thus, to block activation triggered by telomere shortening of cell suicide regulatory and/or execution genes that are presumably located in transcriptionally silent subtelomeric regions of chromosomes, mutations may be required for cell immortalization and/or resistance to apoptosis. In the latter case, it is interesting to note that a recently cloned major human telomere-binding protein contains a COOH-terminal region with strong homology to the DNA-binding repeats of the *myb* protooncogene,[47] and the *myb* protooncogene can accelerate apoptosis induced by TGFβ1.[48] It has also been reported recently that overexpression of *c-myb* protooncogene strongly stimulated p53-mediated activation of the cell death promoter Bax and enhanced apoptosis of human and murine cells.[62] The evidence that both incomplete replication as well as degradation of telomeric DNA may cause telomere loss in metastatic cancer cells further emphasizes the significance of active telomerase in the maintenance of telomere lengths in malignant cells. The direct correlation of the extent of telomeric DNA degradation with the level of apoptosis in metastatic cancer cells suggests that telomerase inhibitors should have a potential as effective antimetastatic agents. Our findings also imply that application of short synthetic telomere-specific oligonucleotides for modification of apoptosis may serve as a conceivable prototype of oligonucleotide-based therapy of apoptosis-related disorders in humans.

In conclusion, here we present three independent lines of experimental evidence strongly indicating that telomere status may determine sensitivity to apoptosis of metastatic cancer cells.

1. Degradation of telomeric DNA is associated with induction of apoptosis. We employed the PCR-based assay with telomere-specific primers for analysis of telomeric repeat degradation products (TRDP) in low molecular weight DNA extracts from nine murine and human cancer cell lines. The TRDP assay revealed an asymmetrical accumulation of (CCCTAA)*n* telomere fragments in proliferating in vitro melanoma and breast carcinoma cells, suggesting a nuclease-mediated mechanism for G-tail formation on chromosome ends, and providing first experimental evidence that such a mechanism exists in mammalian cells. We document that enhanced degradation of telomeric DNA

is associated with increased activity of nuclear Ca^{2+}-dependent endonucleases and higher levels of apoptosis in a panel of cancer cell lines exhibiting different metastatic potentials and different apoptosis levels. Decreased degradation of telomeric DNA, low nuclear endonuclease activity, and diminished levels of apoptosis were associated with high metastatic potential. Induction of apoptosis in metastatic cancer cells by serum deprivation was associated with increased levels of TRDP and required activation of ICE-like protease(s), particularly CPP32/Yama death protease. However, degradation of telomeric DNA was not affected when cells were treated with a specific peptide inhibitor of ICE-like protease(s), indicating that functional ICE-like protease(s) was not required for endonuclease-dependent telomere fragmentation.

2. Telomere-specific oligonucleotides inhibit induction of apoptosis. Treatment of MDA-MB-435 human breast carcinoma cells with short telomere-specific oligonucleotide $(TTAGGG)_3$ significantly inhibits serum deprivation-induced apoptosis. In contrast, treatment with $(CCCTAA)_3$ oligonucleotides has no effect. TTAGGG-induced inhibition of apoptosis does not interfere with CPP32/Yama death protease activation, implying that activation of CPP32/Yama is not sufficient for apoptosis execution in the cells treated with TTAGGG telomere-specific oligonucleotides. Thus, both telomere degradation and CPP32/Yama death protease activation are essential and necessary components of apoptotic cascade.
3. Cells with elongated telomeres exhibit an increased resistance toward apoptosis. The GTR-DU145 human prostate adenocarcinoma cells and GTR-IDH4 immortal human fibroblast cells with stable experimentally elongated telomeres manifest a substantially higher resistance to apoptosis, decreased activity of ICE-like death proteases (caspases) and diminished inducibility of Ca^{2+}-dependent nuclear endonucleases compared to the parental cells DU145 and IDH4, confirming the significance of telomere length in regulation of apoptosis.

Taken together, our data strongly indicate that telomere degradation is closely related to apoptosis, and telomeric DNA repeats are direct targets for apoptosis-associated nuclear Ca^{2+}-dependent endonucleases. Our findings also imply that enhanced degradation of telomeric DNA could generate an intracellular signal(s) triggering apoptosis through amplification of chromosome instability and acceleration of apoptotic degradation of DNA. We therefore suggest that degradation of telomeres is a common intracellular biochemical pathway of replicative cell senescence and apoptosis, and that programmed cell death could be defined as a catastrophic cell senescence.

Acknowledgements

We thank I. Fidler and J. Price (M.D. Anderson Cancer Center, Houston, TX) for providing metastatic cancer cell lines; Dr. J. Shay (UT SW Medical Center, Dallas, TX) for providing cell lines with experimentally elongated telomeres. This work was supported in part by NIH/NCI Grant 1R43CA61402-01 and MetaStat, Inc.

References

1. Blackburn EH, Szostak JW. The molecular structure of centromeres and telomeres. Annu Rev Biochem 1984; 53:163-194.
2. Zakian VA. Structure and function of telomeres. Annu Rev Genet 1989; 23:579-604.
3. Blackburn EH. Structure and function of telomeres. Nature 1991; 350:569-573.
4. Zakian VA. Telomeres: begining to understand the end. Science 1995; 270:1601-1607.
5. Moyzis RK, Buckingham JM, Cram LS, Dani M, Deaven LL, Jones MD, Meyne J, Ratliff RL, Wu JR. A highly conserve, repetitive DNA sequence(TTAGGG)n present at the telomeres of human chromosomes. Proc Natl Acad Sci USA 1988; 85:6622-6626.
6. Meyne J, Ratliff RL, Moyzis RK. Conservation of the human telomere sequence (TTAGGG)n among vertebrates. Pro Natl Acad Sci USA 1989; 86:7049-7053.
7. Muller HJ. The remaking the chromosomes. The Collecting Net 1938; 8:182.
8. McClintock B. The behavior in successive nuclear divisions of a chromosome broken at meiosis. Proc Natl Acad Sci USA 1939; 25:405-416.
9. McClintock B. The stability of broken ends of chromosomes in *Zea mays*. Genetics 1941; 26:234-282.
10. Greider CW, Blackburn EH. Identification of a specific telomere terminal transferase activity in Tetrahymena extracts. Cell 1985; 43:405-412.
11. Yu GL, Bradley JD, Attardi LD, Blackburn EH, Nature 1990; 344:126.
12. Greider CW, Blackburn EH. A telomeric sequence in the RNA of Tetrahymena telomerase required for telomere repeat synthesis. Nature 1989; 337:331.
13. Harley CB, Futcher AB, Greider CW. Telomeres shorten during aging of human fibroblasts. Nature 1990; 345:458-460.

14. Allsopp RC, Vaziri H, Patterson S, Goldstein S, Younglai EV, Futcher AB, Greider CW, Harley CB. Telomere length predicts replicative capacity of human fibroblasts. Proc Natl Acad Sci USA 1992; 89:10114-10118.
15. Vaziri H, Schachter F, Uchida I, Wei L, Zhu X, Effros R, Cohen D, Harley CB. Loss of telomeric DNA during aging of normal and trisomy 21 human lymphocytes. Am J Hum Genet 1993; 52:661-667.
16. Watson JD. Origin of concatemeric T7 DNA. Nature New Biol 1972; 239:197-201.
17. Olovnikov AM. A theory of marginotomy: The incomplete copying of template margin in enzymic synthesis of polynucleotides and biological significance of the phenomenon. J Theor Biol 1973; 41: 181-190.
18. Hastie ND, Dempster M, Dunlop MG, Thompson AM, Green DK, Allshire RC. Telomere reduction in human colorectal carcinoma and with aging. Nature 1990; 346:866-868.
19. Harley CB. Telomere loss: mitotic clock or genetic time bomb? Mutat Res 1991; 256:271-282.
20. Lindsey J, McGill NI, Lindsey LA, Green DK, Cooke HJ. In vivo loss of telomeric repeats with age in humans. Mutat Res 1991; 256:45-48.
21. Counter CM, Avilion AA, LeFeuvre CF, Stewart NG, Greider CW, Harley CB, Bacchetti S. Telomere shortening associated with chromosome instability is arrested in immortal cells which express telomerase activity. EMBO J 1992; 11:19211929.
22. Vaziri H, Dragowska W, Allsopp RC, Thomas TE, Harley CB, Lansdorp PM. Evidence for a mitotic clock in human hematopoietic stem cells: loss of telomeric DNA with age. Proc Natl Acad Sci USA 1994; 91:9857-9860.
23. Wright WE, Shay JW. Telomere positional effects and the regulation of cellular senescence. Trends Genet 1992; 8:193-197.
24. Counter CM, Botelho FM, Wang P, Harley CB, Bacchetti S. Stabilization of short telomeres and telomerase activity accompany immortalization of Epstein-Barr virus-transformed human B-lymphocytes. J Virol 1994; 68:3410-3414.
25. Nilsson P, Mehle C, Remes K, Roos G. Oncogene 1994; 9:3043.
26. Kim NW, Piatyszek MA, Prowse KR, Harley CB, West MD, Ho PL, Coviello GM, Wright WE, Weinrich CL, Shay JW. Specific association of human telomerase activity with immortalized cells and cancer. Science 1994; 266:2011-2015.
27. Counter CM, Hirte HW, Bacchetti S, Harley CB. Telomerase activity in human ovarian carcinoma. Proc Natl Acad Sci USA 1994; 91:2900-2904.
28. Feng J, Funk W, Wang S, Weinrich S, Avilion A, Chiu C-P, Adams R, Chang E, Yu J, Le S, West M, Harley CB, rews W, Greider CW, Villeponteau B. The human telomerase RNA component. Science 1995; 269:1236-1241.
29. De Lange T, Shiue L, Myers RM, Cox DR, Naylor SL, Killery AM, Varmus HE. Structure and variability of human chromosome ends. Mol Cell Biol 1990; 10:518-527.

30. Adamson DJA, King DJ, Haites NE. Significant telomere shortening in childhood leukemia. Cancer Genet Cytogenet 1992; 61:204-205.
31. Smith JK, Yeh G. Telomere reduction in endometrial adenocarcinoma. Am J Obstet Gynecol 1992; 167:1883-1887.
32. Holzmann K et al. Genes Chromosomes Cancer 1993; 6:178.
33. Schwartz HS, Dahir GA, Butler MG. Cancer Genet Cytogenet 1993; 71:132.
34. Hiyama E, Yokoyama T, Tatsumoto N, Hiyama K, Imamura Y, Murakami Y, Kodama T, Piatyszek MA, Shay JW, Matsuura Y. Telomerase activity in gastric cancer. Cancer Res 1995; 55:3258-3262.
35. Counter CM, Gupta J, Harley CB, Leber B, Bacchetti S. Blood 1995; 85:2315.
36. Sommerfeld, H-J, Meeker AK, Piatyszek MA, Bova GS, Shay JW, Coffey DS. Telomerase activity: a prevalent marker of malignant human prostate tissue. Cancer Res 1996; 56:218-222.
37. Levy MZ, Allsopp RC, Futcher AB, Greider CW, Harley CB. Telomere end-replication problem and cell aging. J Mol Biol 1992; 225:951-960.
38. Wyllie AH, Kerr JFR, Currie AR. Cell death: The significance of apoptosis. Int Rev Cytol 1980; 68:251-306.
39. Thompson CB. Apoptosis in the pathogenesis and treatment of disease. Science 1995; 267:1456-1462.
40. Williams GT, Smith CA. Molecular regulation on apoptosis: Genetic control on cell death. Cell 1993; 74:777-779.
41. Raff MC. Social control on cell survival and cell death. Nature 1992; 356:397-400.
42. Williams GT. Programmed cell death: Apoptosis and oncogenesis. Cell 1991; 65:1097-1098.
43. Reed JC. Bcl-2 and the regulation of programmed cell death. J Cell Biol 1994; 124:1-6.
44. Glinsky GV, Glinsky VV. Apoptosis and metastasis: A superior resistance of metastatic cancer cells to programmed cell death. Cancer Letters 1996; 101:43-51.
45. Wyllie AH. Glucocorticoid-induced thymocyte apoptosis is associated with endogenous endonuclease activation. Nature 1980; 284:555-556.
46. McDonell MW, Simon MN, Studier FW. Analysis of restriction fragments of T7 DNA and determination of molecular weights by electrophoresis in neutral and alkaline gels. J Mol Biol 1977; 110:119.
47. Chong L, van Steensel B, Broccoli D, Erdjument-Bromage H, Hanish J, Tempst P, De Lange T. A human telomeric protein. Science 1995; 270:1663-1667.
48. Selvakumaran M, Lin HK, Tjin Tham Sjin R, Reed JC, Liebermann DA, Hoffman B. Mol Cell Biol 1994; 14:2352-2360.
49. Wright WE, Brasiskyte D, Piatyszek MA, Shay JW. Experimental elongation of telomeres extends the lifespan of immortal x normal cell hybrids. EMBO J 1996; 15:1734-1741.

50. Zhang RD, Fidler IJ, Price JE. Relative malignant potential of human breast carcinoma cell lines established from pleural effusions and a brain metastasis. Invasion & Metastasis 1991; 11:204-215.
51. Glinsky GV, Mossine VV, Price JE, Bielenberg D, Glinsky VV, Ananthaswamy HN, Feather MS. Inhibition of colony formation in agarose of metastatic human breast carcinoma and melanoma cells by synthetic glycoamine analogs. Clin & Exper Metastasis 1996; 14:253-267.
52. Price JE, Polyzos A, Zhang RD, Daniels LM. Tumorigenicity and metastasis of human breast carcinoma cell lines in nude mice. Cancer Res 1990; 50:717-721.
53. Schackert G, Price JE, Zhang R, Bucana CD, Itoh K, Fidler IJ. Regional growth of different human melanomas as metastases in the brain of nude mice. Am J Pathol 1990; 136:95-102.
54. Zhang RD, Price JE, Schackert G, Itoh K, Fidler IJ. Malignant potential of cells isolated from lymph node or brain metastases of melanoma patients and implication for prognosis. Cancer Res 1991; 51:2029-2035.
55. Fidler IJ. The relationship of embolic homogeneity, number, size and viability to the incidence of experimental metastasis. Eur J Cancer 1973; 9:223-227.
56. Fidler IJ. Selection of successive tumor lines for metastasis. Nat New Biol 1973; 242:148-149.
57. Fidler IJ. Biological behavior of malignant melanoma cells correlated to their survival in vivo. Cancer Res 1975; 35:218-224.
58. Hart IR. The selection and characterization of an invasive variant of the B16 melanoma. Am J Pathol 1979; 97:587-592.
59. Pronk GJ, Ramer K, Amiri P, Williams LT. Requirement of an ICE-like protease for induction of apoptosis and ceramide generation by REAPER. Science 1996; 271:808-810.
60. Arends MJ, McGregor AH, Toft NJ, Brown EJH, Wyllie AH. Susceptibility to apoptosis is differentially regulated by c-myc and mutated Ha-ras oncogenes and is associated with endonuclease availability. Br J Cancer 1993; 68:1127-1133.
61. McConkey DJ, Chandra J, Wright S, Plunkett W, McDonnell TJ, Reed JC, Keating MJ. Apoptosis sensitivity in chronic lymphocytic leukemia is determined by endogenous endonuclease content and relative expression of Bcl-2 and Bax. J Immunol 1996; 156:2624-2630.
62. Sala A, Casella I, Grasso L, Bellon T, Reed JC, Miyashita T, Peschle C. Apoptosis response to oncogenic stimuli: Cooperative and antagonistic interactions between c-myb and the growth suppressor p53. Cancer Res 1996; 56:1991-1996.
63. McEachern MJ, Blackburn EH. Runaway telomere elongation caused by telomerase RNA gene mutations. Nature 1995; 376:403-409.
64. Lingner J, Cooper JP, Cech TR. Telomerase and DNA end replication: no longer a lagging strand problem? Science 1995; 269: 1533-1534.

65. Wellinger RJ, Wolf AJ, Zakian VA. Origin activation and formation of single strand TG1-3 tails occur sequentially in the late S phase on a yeast linear plasmid. Mol Cell Biol 1993; 13:4057-4065.
66. Wellinger RJ, Wolf AJ, Zakian VA. Saccharomyces telomeres acquire single-strand TG1-3 tails late in S phase. Cell 1993; 72:51-60.
67. Wellinger RJ, Ethier K, Labrecque P, Zakian VA. Evidence for a new step in telomere maintenance. Cell 1996; 85:423-433.
68. Harrington JJ, Lieber MR. Functional domains within FEN-1 and RAD2 define a family of structure-specific endonucleases: implication for nucleotide excision repair. Genes Dev 1994; 8:1344-1355.
69. Glinsky GV, Price JE, Glinsky VV, Mossine VV, Kiriakova G, Metcalf JB. Inhibition of human breast cancer metastasis in nude mice by synthetic glycoamines. Cancer Res 1996; 56:5319-5324.
70. Hiyama E, Hiyama K, Yokoyama T, Ichikawa T, Matsuura Y. Length of telomeric repeats in neuroblastoma: correlation with prognosis and other biological characteristics. Jpn J Cancer Res 1992; 83:159-164.
71. Mehle C, Ljungberg B, Roos G. Telomere shortening in renal cell carcinoma. Cancer Res 1994; 54:236-241.
72. Fernandes-Alnemri T, Litwack G, Alnemri ES. CPP32, a novel human apoptotic protein with homology to Caenorhabditis elegans cell death protein Ced-3 and mammalian interleukin-1β-converting enzyme. J Biol Chem 1994; 269:30761-30764.
73. Nicholson DW, Ali A, Thornberry NA, Vaillancourt JP, Ding CK, Gallant M et al. Identification and inhibition of the ICE/CED-3 protease necessary for mammalian apoptosis. Nature 1995; 376:37-39.
74. Tewari M, Quan LT, O'Rourke K, Desnoyers S, Zeng Z, Beidler DR, Poirier GG, Salvesen GS, Dixit VM. Yama/CPP32b, a mammalian homolog of CED-3, is a crm-inhibitable protease that cleaves the death substrate poly (ADP-ribose) polymerase. Cell 1995; 81:801-809.
75. Lazebnik YA, Kaufmann SH, Desnoyers S, Poirier GG, Earnshaw WC. Cleavage of poly (ADP-ribose) plymerase by a proreinase with properties like ICE. Nature 1994; 371:346-347.
76. Casciola-Rosen LA, Miller DK, Anhalt GJ, Rosen A. Specific cleavage of the 70 kDa protein component of small nuclear ribonucleoprotein is a characteristic biochemical feature of apoptotic cell death. J Biol Chem 1994; 269:30757-30760.
77. Thornberry NA. Interleukin-1β-converting enzyme. Methods Enzymology 1994; 244:615-631.
78. Glinsky GV, Glinsky VV, Ivanova AB, Hueser CN. Apoptosis and Metastatsis: increased apoptosis resistance of metastatic cancer cells is associated with the profound deficiency of apoptosis execution mechanisms. Cancer Lett 1997; 115:185-193.
79. Glinsky GV. Apoptosis in metastatic cancer cells. Crit Rev Oncol/Hemat 1997; 25:175-186.
80. Holt S, Glinsky VV, Shay J, Glinsky GV. Resistance to apoptosis is associated with telomere length and telomerase function. Submitted.

CHAPTER 5

Inhibition of Homotypic Cancer Cell Aggregation and Colony Formation in Agarose of Metastatic Human Cancer Cells by Synthetic Glycoamine Analogs

5.1. Summary

The following two chapters will be significantly different from the previous ones. They will describe our recent experimental efforts in practical implementation of theoretical considerations outlined in previous chapters: development of a new family of anti-metastatic drugs acting on the molecular level as inhibitors of cancer cell adhesion and inducers of apoptosis in metastatic cancer cells. In this chapter we will describe our initial in vitro screening strategy which will ultimately lead to the selection of several potential anti-metastatic agents called anti-metastatic synthetic glycoamines. We studied the influence of ten synthetic glycoamine analogs on colony formation in 0.3% and 0.9% agarose by metastatic human breast carcinoma (MDA-MB-435) and melanoma (TXM-13) cells. Nine synthetic analogs significantly inhibited the colony formation in 0.9% agarose of MDA-MB-435 human breast carcinoma cells; five compounds caused a 73%-83% reduction of colony formation. Seven synthetic glycoamines caused a significant inhibition of colony formation in 0.9% agarose by TXM-13 melanoma cells, with the inhibitory effect ranging from 71%-87%. The 50% inhibition (I_{50}) doses and relative activity rank of the compounds were similar for both breast carcinoma and melanoma cell lines. The murine B16 mela-

Failure of Apoptosis and Cancer Metastasis, by Gennadi V. Glinsky.

noma cell aggregation assay was employed to elucidate the potential mechanism(s) of the inhibitory activity of synthetic glycoamines. The relative activity ranks of the compounds based on the independently determined I_{50} doses for both cell aggregation and clonogenic growth assays were very similar for the four most active synthetic analogs and clearly indicated the importance of hydrophobic amino acids in mediating the bioactivity of synthetic glycoamines. In both experimental systems (clonogenic growth in agarose and cell aggregation assay), the leading compound was N-(1-deoxy-D-fructos-1-yl)-D-leucine (Fru-D-Leu) and the least active analog was N-(1-deoxy-D-fructos-1-yl)-glycine (Fru-Gly). These results show that synthetic glycoamines may act by competing for specific carbohydrate-lectin interactions, particularly those involving β-galactoside-specific lectins expressed on metastatic cells.

5.2. Introduction

Metastasis is the most important malignant feature of human cancer and represents the morphological stage of the generalization of the disease through the body of the tumor host. Since metastasis is a major cause of death of cancer patients, there is an obvious clinical need for effective antimetastatic drug(s). In vitro screening and evaluation of potential therapeutic compounds, as well as establishing the cellular and molecular mechanisms of their action, are important steps in the development of antimetastatic drug(s).

The anchorage-independent growth of cells in a semisolid medium is considered an important biological property closely associated with malignant transformation and tumorigenicity.[1] The clonogenic growth assay has been used for a variety of investigations into the growth potential and hormone and drug sensitivity of surgical tumor specimens.[2,3] The clonogenic growth potential in agarose was found to correlate with the progression of breast neoplasms from benign to malignant.[4] Colony formation in agar cultures was found to be an independent variable associated with shorter disease-free survival and overall survival after mastectomy.[5] In many experimental systems, colony formation in agar was shown to be correlated with metastatic potential.[6,7] A particularly discriminating assay was to test colony forming efficiency in higher density concentrations of agar medium.[8,9] It has been shown that for human tumor cell lines the ability to grow in dense agarose cultures (>0.6% agarose) also corresponds well with the experimental metastatic ability, tested in nude mice.[9-12] Thus, colony formation in cultures of agarose greater than 0.6% distinguishes tumor cells with enhanced metastatic potential[10-12] and, therefore, the clonogenic growth assay could be used for in vitro evaluation of potential antimetastatic compounds, particularly for therapy of metastatic breast cancer and melanoma.

Early interest in glycoamines as naturally occurring low molecular weight components of mammalian blood serum was aimed at evaluating their role as potential humoral cancer markers.[13-15] The prototype compounds were originally isolated from human and rodent blood serum employing a combination of conventional chromatography, analytical ultrafiltration, and size-exclusion and RP HPLC,[13-18] and subsequently structurally characterized as a glycoconjugate of amino acids with covalent structures of Schiff base, ester, and Amadori bonding.[13,15-18] The inhibitory activity of HPLC-purified natural and synthetic glycoamines on cancer cell aggregation in vitro, and experimental metastasis in vivo, was reported primarily rodent cells.[13,14,19-23] The influence of synthetic glycoamine analogs on human cancer cells has not been previously investigated.

Recently we reported a method of synthesis, purification and full structural characterization of a series of synthetic glycoamine analogs, using NMR, MS and X-ray analysis.[24,25] The availability of synthetic glycoamine analogs in pure form in multigram quantities would allow extensive in vitro and in vivo evaluation of the anti-adhesion and anti-metastatic activity of synthetic glycoamines. The main objective of this study was to investigate the influence of ten synthetic glycoamine analogs on the clonogenic growth in hard agarose of metastatic human breast carcinoma and melanoma cell lines. In this paper we would like to report a simple procedure for the synthesis, purification and isolation in crystalline form of six glycoamine analogs and present their structural characteristics, including FAB MS and NMR data. We also document the significant (over 80%) inhibition of colony formation in hard agarose of human breast carcinoma and melanoma cell lines that are metastatic in nude mice, by a series of synthetic glycoamine analogs, particularly by Amadori compounds of hydrophobic amino acids. Experimental evidence will be considered which suggests that carbohydrate-based adhesion of tumor cells to the polygalactoside matrices may play a significant role in the ability of malignant cells to grow in semisolid medium.

5.3. Structural Analysis of Synthetic Glycoamine Analogs

Structural formulae and structural analysis data for synthetic compounds tested in this work are presented in Table 5.1 and Figure 5.1.[74] The results of the structural analysis including FAB MS and NMR data are in good agreement with previously established criteria of purity and structural confirmation for this class of compounds (ref. 24 and references therein). In aqueous solutions, Amadori compounds reveal conformational instability of their sugar part and exist, according to NMR data, in four major tautomeric forms: α- and β-pyranoses and furanoses (Table 5.1; Fig. 5.2). The population of the open chain form is small (less than 1%) and can be detected by NMR technique only with extensive scanning. Thus, this form may be excluded from the structure-activity considerations. Any of the four major conformations may posses specific biologic activity. However, they undergo mutual interconversion quickly and the equilibrium between them may be achieved in 2-3 min. The distribution of any Amadori compounds between tautomeric forms is essentially the same (Table 5.1) and does not depend on the nature of amino acid or carbohydrate side chains. Thus, the tautomeric population does not appear to vary significantly with changes in the amine residue, and, for Amadori compounds prepared from D-glucose and a number of amino acids, the population is about 60-70% β-pyranose, about 20-30% α- and β-furanose, and about 5% α-pyranose. The established difference in activities of compounds having a common sugar moiety (fructose in 1-9) or amino acid part (L-leucine in 7 and 10) indicates that factors other than sugar tautomeric form distribution may play a major role in determining the bioactivity of the Amadori compounds. Thus, the Amadori compounds may interact with cell membrane receptor(s) as more than monodentate ligands, since both carbohydrate and amino acid constituents play a role in inhibition of cancer cell adhesion and clonogenic growth (see below), and neither fructose nor free amino acids possess similar antiadhesion activity.

5.4. Inhibition of Colony Formation in Agarose by Synthetic Glycoamines

Ten selected synthetic structural glycoamine analogs were tested in colony formation assay using human melanoma (TXM-13) and breast carcinoma (MDA-MB-435) cells. Nine compounds significantly inhibited colony formation of the MDA-MB-435 cells at concentrations from 0.375 to 6.0 mM (Fig. 5.3). Four of these compounds

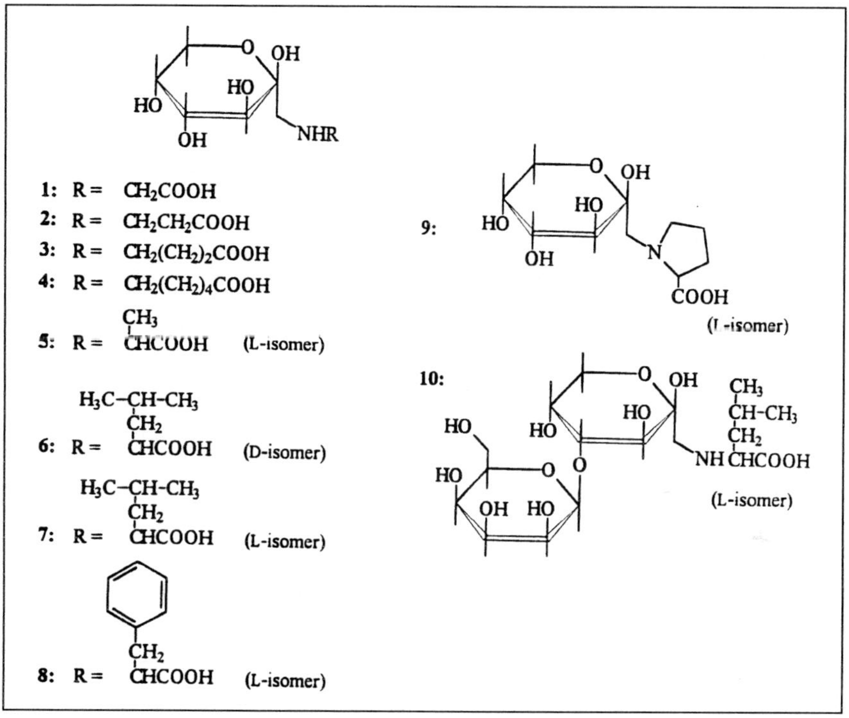

Fig. 5.1. Structural formulae of the synthetic glycoamines tested in this work. Reproduced with permission from Glinsky GV et al, Clin Exper Metastasis 1996; 14:253-267.

Table 5.1. An equilibrium composition (%) of tautomeric forms of Amadori compounds in D_2 (solutions estimated from ^{13}C-NMR spectra

	1	2	3	4	5	6	7	8	9	10
β-pyranose	69	70	70	72	72	71	72	69	67	65
α-pyranose	5	4	5	5	5	4	4	5	4	3
β-furanose	13	11	13	11	12	11	12	11	15	15
α-furanose	13	15	12	12	11	14	12	15	14	17

produced inhibition ranging from 42% to 63%, and 5 compounds caused a reduction in colony formation in agarose ranging from 73% to 83% (Table 5.2).

Seven synthetic compounds caused a significant inhibition of colony formation by TXM-13 melanoma cells in 0.9% agarose, producing from 71 to 87% inhibition (Fig. 5.3; Table 5.2). For most of the

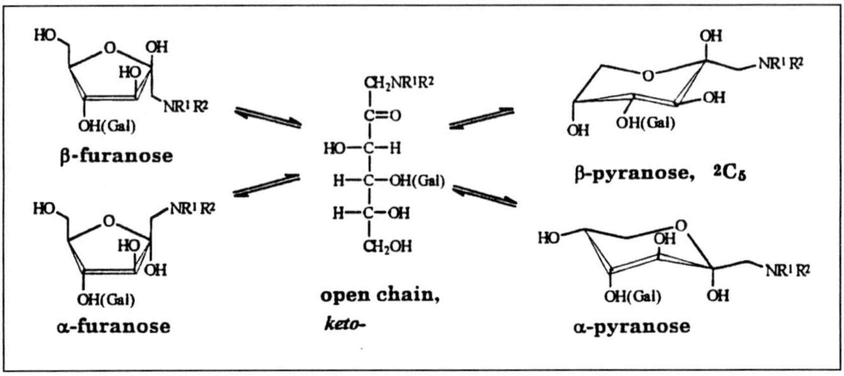

Fig. 5.2. Tautomeric forms of the Amadori compounds in an aqueous solutions. Reproduced with permission from Glinsky GV et al, Clin Exper Metastasis 1996; 14:253-267.

Table 5.2. The maximum inhibitory activity (percent of inhibition) of synthetic glycoamines in colony formation in agarose assay

Compounds	MDA-MB-435 Human Breast Carcinoma		TXM-13 Human Melanoma	
	Concentration of Agarose			
	0.9%	0.3%	0.9%	0.3%
Fru-β-Ala(SSGA-1)	78.3	40.5	NT[a]	NT
Fru - Gly(SSGA-2)	33.8[b]	47.3	47.3	73.1
Fru-L-Phe(SSGA-3)	83.0	5.4[b]	78.3	84.0
Fru-γ-Abu(SSGA-9)	NT	29.5[b]	NS[b]	NS[b]
Fru-ε-Aca(SSGA-10)	56.9	16.5[b]	NS[b]	NS[b]
Fru-L-Leu(SSGA-12)	51.3	62.9	56.2	76.7
Fru-D-Leu(SSGA-13)	74.2	2.3[b]	86.9	79.3
Fru-L-Ala(SSGA-14)	41.9	31.2[b]	68.7	71.2
Fru-L-Pro(SSGA-16	72.7	67.5	72.2	58.6
Lac-L-Leu(SSGA-19)	82.3	41.3	74.4	69.1

[a](NT), not tested; [b], The difference is not statistically significant.

Fig. 5.3 (opposite). Dose-response analysis of the inhibition of colony formation in 0.9% agarose cultures of human breast carcinoma MDA-MB-435 (top three panels) and human melanoma TXM-13 (bottom three panels) cell lines by synthetic glycoamine analogs. Key: SSGA-3: Fru-L-Phe; SSGA-12: Fru-L-Leu; SSGA-13: Fru-D-Leu; SSGA-19: Lac-L-Leu). Single cell suspension of cells were mixed with agarose and plated in 35 mm dishes with or without the addition of tested compounds at given concentration. Colonies were counted at 14 days (0.3% agarose) or 21-25 days (0.9% agarose). The values shown are the mean number of colonies of diameter greater than 50 μm (±SEM) in 3 replicate dishes with size of colonies measured using an eyepiece micrometer. The number of colonies was counted in a sample area of the dish (1 cm^2) and the number in the total area of the 35 mm diameter dish calculated (multiplication by 9.62). Reproduced with permission from Glinsky GV et al, Clin Exper Metastasis 1996; 14:253-267.

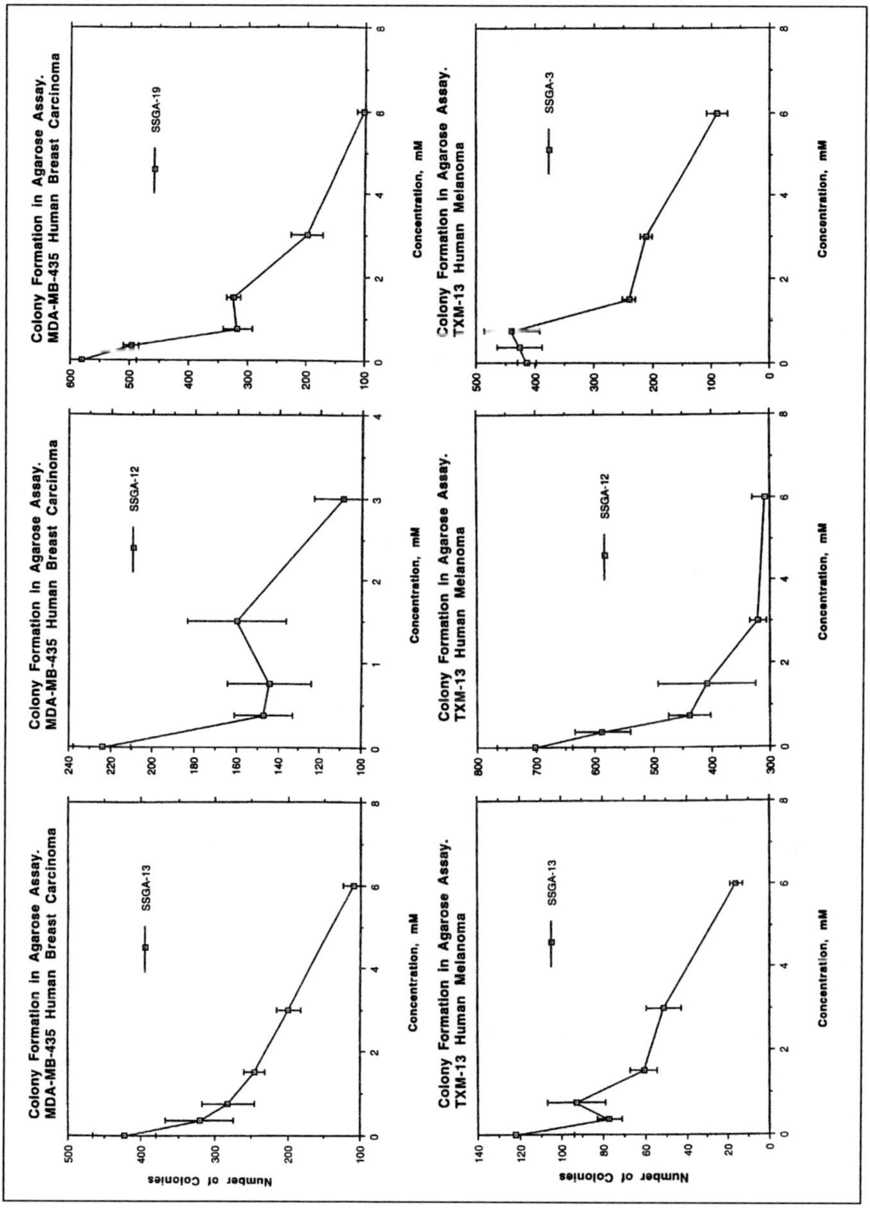
Colony Formation in Agarose Assay. MDA-MB-435 Human Breast Carcinoma
SSGA-13
Colony Formation in Agarose Assay. MDA-MB-435 Human Breast Carcinoma
SSGA-12
Colony Formation in Agarose Assay. MDA-MB-435 Human Breast Carcinoma
SSGA-19
Colony Formation in Agarose Assay. TXM-13 Human Melanoma
SSGA-13
Colony Formation in Agarose Assay. TXM-13 Human Melanoma
SSGA-12
Colony Formation in Agarose Assay. TXM-13 Human Melanoma
SSGA-3
Number of Colonies
Concentration, mM

compounds greater inhibition of colony formation by the human cells was seen in cultures of dense agarose (0.9%) than in 0.3% agarose cultures. Two compounds (SSGA-9 and SSGA-10) produced no significant inhibition of colony formation by either TXM-13 or MDA-MB-435 cells in 0.3% agarose cultures (Table 5.2)When the active compounds were ranked in terms of I_{50} doses, four compounds were found to be the most effective for the inhibition of colony formation by both human cell lines (Table 5.3). One exception was Fru-Gly (SSGA-2) which was active against TXM-13 cells yet produced no significant inhibition of MDA-MB-435 colony formation (Tables 5.2, 5.3).

Therefore, we have identified a number of compounds that were effective inhibitors of colony formation in dense agarose by human melanoma and breast carcinoma cells.

5.5. Inhibition of B16 Melanoma Cell Aggregation by Synthetic Glycoamines

It is well established for a B16 melanoma experimental metastasis model in mice that galactoside-specific lectins and corresponding carbohydrate-protein recognition and binding play a key role in melanoma cell aggregation in vitro and metastasis in vivo (see Conclusion). We have employed this model for studying the mechanism(s) of the inhibitory action of synthetic glycoamines on cancer cell aggregation, clonogenic growth and metastasis. A panel of 28 synthetic glycoamine analogs were tested against B16-F10 cells in an aggregation assay. Five compounds were identified as strong inhibitors of cell aggregation (43-80% inhibition) (Table 5.4, Fig. 5.4). Two of these compounds were more effective against the highly metastatic B16-F10 cells than the poorly metastatic B16-F1 cells (Table 5.5). The relative activity ranking of the compounds in terms of I_{50} doses from the cell aggregation and clonogenic growth assays were similar for the four most biologically active synthetic analogs (Tables 5.3 and 5.5), which may indicate some similarity in the mechanism of action in the different assays. In both assays, testing inhibition of aggregation of B16 melanoma cells or inhibition of colony formation by human melanoma or breast carcinoma cells, the most active compound was Fru-D-Leu (SSGA-13). No cytotoxicity was observed based on trypan blue dye exclusion assay following up to 96 h incubation of B16-F10 melanoma cells with 5 mM Fru-D-Leu.

Table 5.3. The 50% inhibition doses of synthetic glycoamines in colony formation in agarose assay

Compounds	The 50% Inhibition Dose (mM) in 0.9% Agarose	
	MDA-MB-435 Human Breast Carcinoma	TXM-13 Human Melanoma
Lac-L-Leu(SSGA-19)	2.3 mM	1.1 mM
Fru-D-Leu(SSGA-13)	2.3 mM	1.5 mM
Fru-L-Leu(SSGA-12)	3.0 mM	2.3 mM
Fru-L-Phe(SSGA-3)	2.7 mM	3.0 mM
Fru-β-Ala(SSGA-1)	3.8 mM	NT[a]
Fru-ε-Aca(SSGA-10)	5.3 mM	NS[b]
Fru-L-Pro(SSGA-16)	5.3 mM	3.8 mM
Fru-L-Ala(SSGA-14)	>6.0 mM	3.0 mM
Fru - Gly(SSGA-2)	NS	1.5 mM
Fru-γ-Abu(SSGA-9)	NT	NS

[a](NT), not tested; [b](NS), the difference is not statistically significant.

Table 5.4. The maximum inhibitory activity (percent of inhibition) of synthetic glycoamines in cell aggregation assay employing poorly (B16 F1) and highly (B16 F10) metastatic murine melanoma cell lines

Compounds	B16 Fl Melanoma	B16 F10 Melanoma
Fru-L-Phe(SSGA-3)	44%[a]	48%[a]
Fru-L-Leu(SSGA-12)	43%	59%
Fru-D-Leu(SSGA-13)	34%	80%
Lac-L-Leu(SSGA-19)	78%	43%

a The difference in activity between B16 F10 and B16-Fl cells for SSGA-3 analog is not statistically significant.

5.6. Galactoside-Specific Lectins as Potential Molecular Targets for Synthetic Glycoamines

We employed the multicell aggregate formation assay with murine B16-F1 melanoma cells and the established cell aggregation promoter asialofetuin (ASF) to test the hypothesis that β-galactoside-specific lectins are the potential molecular targets for synthetic

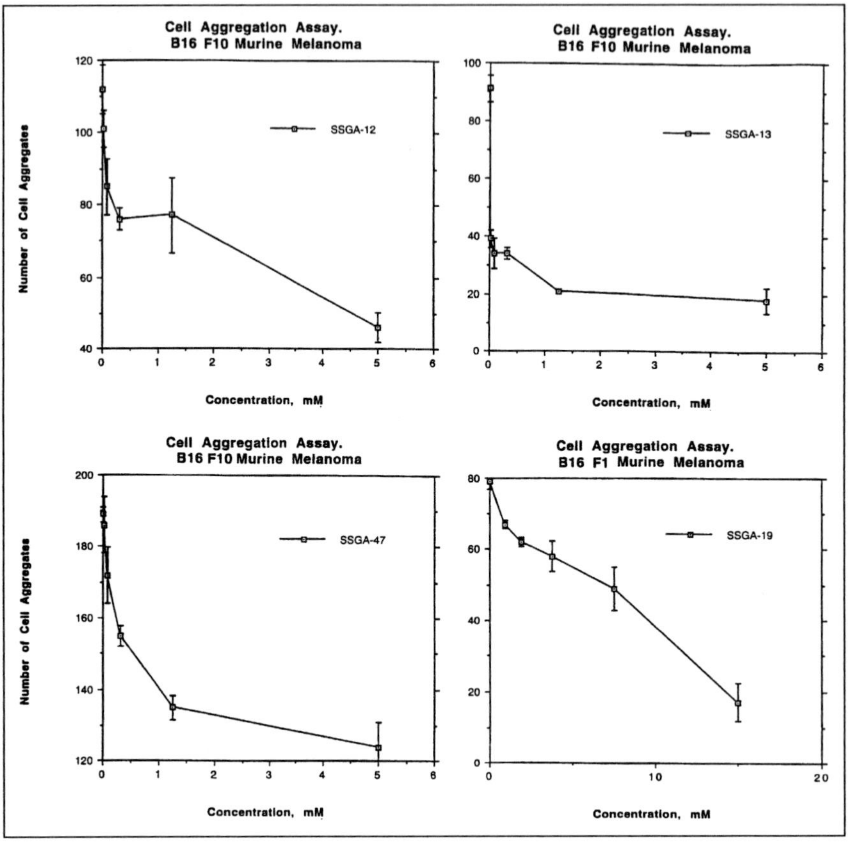

Fig. 5.4. Dose-response analysis of the inhibition of B16-F10 murine metastatic melanoma cell aggregation by synthetic glycoamine analogs. Key: SSGA-12: Fru-L-Leu; SSGA-13: Fru-D-Leu; SSGA-19: Lac-L-Leu; SSGA-47: Fru-L-hydroxyproline. The monocellular suspension of B16-F10 or B16-F1 melanoma cells was cultivated with or without the addition of the tested compounds. The multicellular aggregates were counted after 24 h incubation at 37°C in 5% CO_2. The values shown are the mean number of cell aggregates containing more than 5 cells (±SEM) in 3 replicate wells. The number of cell aggregates was counted using an inverted microscope (magnification, x10) in three random sample areas of each well and the mean number was calculated. Reproduced with permission from Glinsky GV et al, Clin Exper Metastasis 1996; 14:253-267.

glycoamines. The choice of the experimental model was based on the following considerations. It is well established that ASF displays multiple βGal$^{1\text{-}3}$GalNac and βGal$^{1\text{-}4}$GlcNac sequences. It has also been shown that ASF has the ability to increase B16 melanoma cell aggregation in vitro and metastasis in vivo through interaction with β-galactoside-specific lectins that are overexpressed on metastatic cancer cells (see Conclusion). Indeed, the addition of ASF signifi-

Table 5.5. The 50% inhibition doses of synthetic glycoamines in cell aggregation assay employing the highly metastatic B16 F10 murine melanoma cell line

Compounds	Inhibition Dose (mM)
Fru-D-Leu(SSGA-13)	0.02 mM
Fru-L-Trp(SSGA-11)	0.31 mM
Fru-L-Leu(SSGA-12)	3.1 mM
Fru-L-Phe(SSGA-3)	5.0 mM
Fru-L-Hyp(SSGA-47)	>5.0 mM
Lac-L-Leu(SSGA-19)	9.7 mM[a]

a The maximum inhibitory activity has been detected using the B16 F1 murine melanoma cell line.

cantly increased homotypic cell aggregation of B16-F1 melanoma cells (Fig. 5.5A). The ASF-mediated B16-F1 melanoma cell aggregation was almost completely abolished by the most biologically active synthetic glycoamine analog, Fru-D-Leu. In contrast, the biologically nonactive analog Fru-Gly did not affect the ASF-mediated B16-F1 melanoma cell aggregation (Fig. 5.5A). The inhibitory effect of Fru-D-Leu is likely to be β-galactoside-specific, since Fru-D-Leu did not influence the β-galactoside-independent BSA-mediated B16-F1 melanoma cell aggregation. Furthermore, Fru-D-Leu directly inhibited the binding of β-galactoside-specific PNA lectin to ASF (Fig. 5.5B), in contrast to Fru-Gly which was ineffective in the same assay.

5.7. Conclusion: Carbohydrate-Lectin Recognition as a Common Molecular Determinant of Cell Aggregation, Clonogenic Growth and Apoptosis

It is well established that homotypic and heterotypic cancer cell adhesion and aggregation is one of the key determinants of the postintravasation stage of metastasis. Many studies indicate that cell surface components play an important role in metastasis,[33-35] in particular in the formation of tumor cell aggregates and emboli. Circulating metastatic tumor cells tend to adhere to other tumor cells (homotypic adhesion) or to host cells including platelets and lymphocytes (heterotypic adhesion) to form emboli, the formation of which is correlated with metastasis.[27,36,37] The lodgment, attachment, and growth of blood-borne neoplastic cells appear to depend on embolization.[38,39] In the B16 melanoma model of metastasis[28,47]

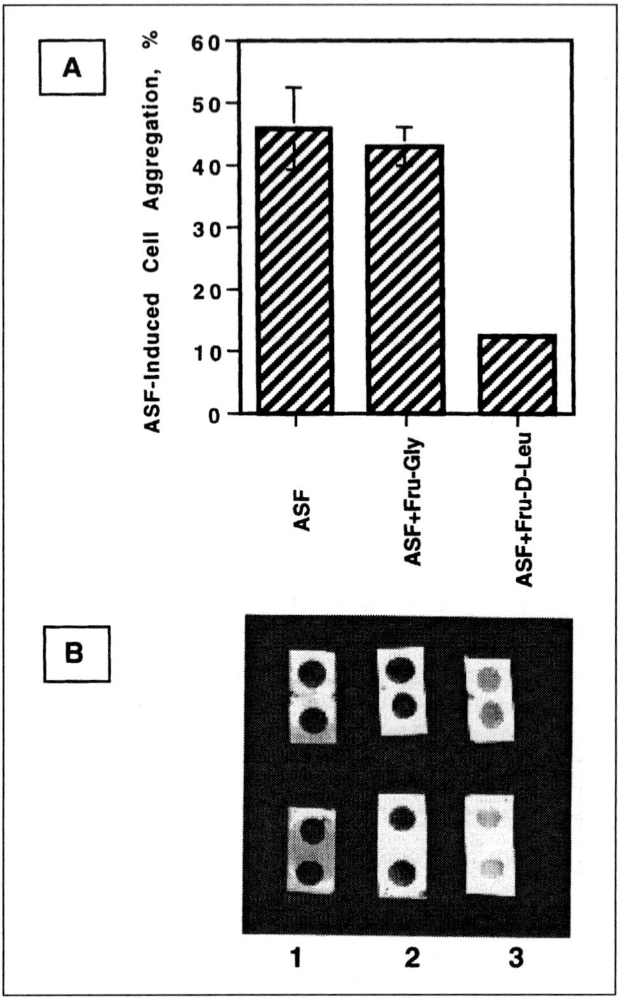

Fig. 5.5. Inhibition by synthetic glycoamines of the ASF-induced B16-F1 melanoma cell aggregation (A) and binding of β-galactoside-specific PNA lectin to ASF (B). A. The monocellular suspension (0.5×10^6 ml^{-1}) of B16-F1 melanoma cells in serum free RPMI 1640 medium with and without addition of 0.25 mg ml^{-1} of ASF (final concentration) and tested compounds were placed in 15 ml plastic cell culture tubes and agitated at 80 rpm for 1 h at 37°C. After 1 h incubation, the homotypic aggregation was determined. The percentage of cells in aggregates of 3 cells or more is presented. The values shown are the mean number (±SD) determined in at least 3 replicate samples. All culture retained > 90% viability throughout the duration of the experiments.
B. The binding of the β-galactoside-specific PNA lectin to the reference glycoprotein ASF displaying βGal as a terminal sugar residue was determined. The glycoproteins (1 γ per dot) were immobilized on the nitrocellulose membranes, blocked with 5% BSA and incubated with peroxidase-labeled peanut lectin (HRP-PNA) alone or with addition of 3 mM (final concentration) of synthetic glycoamines. After 1 h of incubation at room temperature the membranes were washed twice in TBS and developed with DBA. Top panel: Lane 1: PNA, 6.25 γ/ml; Lane 2: PNA + Fru-Gly; Lane 3: PNA + Fru-D-Leu. Bottom panel: Lane 1: PNA, 12.5 γ/ml; Lane 2: PNA, 6.25 γ/ml; Lane 3: PNA, 3.125 γ/ml. Reproduced with permission from Glinsky GV et al, Clin Exper Metastasis 1996; 14:253-267.

it has been shown that tumor cell clumps produce more lung metastases after i.v. injection than do single cells,[27] and there is a correlation between the tendency of cells to undergo homotypic and heterotypic aggregation in vitro and their metastatic potential in vivo.[39-42,48] Tumor cell surface lectins play a major role in in vivo cellular interactions that are important for the formation of emboli and for the arrest of circulating tumor cells.[31,43-46] Tumor cells and tissues contain lectins that are similar to those isolated from normal cells in their sugar-binding specificities, molecular weights, and sequences.[46,49-57] Using monoclonal and polyclonal antilectin antibodies it has been shown that the level of cell surface lectin expression can correlate with malignant transformation and metastatic propensity.[53,54,56] A key role has been suggested for the tumor-surface lectins in anchorage-independent growth in vitro and in tumor embolization (homotypic and heterotypic aggregation) during the metastatic process.[31,43-46] Moreover, galactoside-specific lectins could facilitate the clonogenic growth of metastatic cancer cells in a semisolid medium. The ability to grow in agar or agarose may be influenced by the presence of galactoside-specific lectins on the tumor cell surfaces.[54] In the presence of antibodies to the endogenous lectins, mouse tumor cells show decreased colony-forming efficiency in agar and significantly reduced ability to form lung metastases after i.v. injection in syngeneic mice.[31] Our results support this concept and shed some new light on the mechanism(s) of anchorage-independence of growth of the metastatic cancer cells.

The I_{50} doses of synthetic glycoamine analogs are in the high µM to low mM range, which is typical for monovalent carbohydrate ligands of mammalian lectins.[58] The most active synthetic analog tested so far (Fru-D-Leu) specifically inhibits the binding of both βD-Gal and T-antigen specific carbohydrate biotinylated probes to the B16 F10 murine metastatic melanoma cells (Glinsky et al, unpublished observations), and βD-Gal is a terminal residue on the nonreducing end of T-antigen.[59] Thus, Fru-D-Leu may mimic essential spatial structural features of T-antigen and bind to the T-antigen specific and/or galactoside-specific lectins on the surfaces of metastatic cancer cells, thus disrupting the carbohydrate-protein recognition and binding. Alternatively, it may bind to cell surface receptors for Amadori compounds[32] and indirectly affect the level of expression and/or affinity of cell surface lectins. In this context, it is noteworthy that the overexpression of both T-antigen and T-antigen specific lectins has been observed on about 80-90% of human carcinomas of different origins,[59-61] and T-antigen mediated adhesion is involved in metastatic cell-host cell interactions.[62] Our data also clearly indicate the importance of hydrophobic interactions in facilitating the Amadori compound-mediated inhibition of cancer cell aggregation in vitro and clonogenic growth in a semisolid medium. This is particularly evident in the cell aggregation assay (Tables 5.4, 5.5) as well as the clonogenic growth assay employing MDA-MB-435 cells (Table 5.2): when hydrophobic amino acids such as Leu or Phe in synthetic analogs SSGA-13, SSGA-12, and SSGA-3, which exhibited strong inhibitory activities, were substituted for the Gly residue, the resulting compound SSGA-2 (Fru-Gly) did not manifest a significant biological effect. On the other hand, the least stable synthetic glycoamine analogs such as SSGA-9 (Fru-γ-Abu) and SSGA-10 (Fru-ε-Aca) did not reveal high biological activities, either in cell aggregation or clonogenic growth assays, indicating the importance of the chemical stability of Amadori compounds in preserving their biological activities. The addition of α βGal residue to the Fru-L-Leu (SSGA-12) molecule resulted in 1.3- to 2.1-fold increase in the inhibitory activities of the Lac-L-Leu (SSGA-19) analog, as determined in the clonogenic growth assay in 0.9 % agarose with both MDA-MB-435 and TXM-13 cell lines (Table 5.2). These data further emphasize the significance of β-galactoside-specific interactions in mediating the inhibitory effect of synthetic glycoamine analogs on colony formation in agarose by metastatic cancer cells. Since clusterization of the carbohydrate ligands to bi- or trivalency dramatically increases (from 2,000- to 100,000-fold) their binding affinity,[63] the next logical step in the rational design of more potent glycoamine analogs would include at least bivalency of a carbohydrate constituent combined with a hydrophobic amino acid valency in newly synthesized compounds.

The extended structure-activity study employing over 30 synthetic glycoamine analogs and in vitro cancer cell aggregation assay showed the significance of carbohydrate constituents of glycoamine molecules in mediating the inhibition of homotypic cancer cell aggregation. Several substitutions of the fructosyl residue in Amadori compounds of hydrophobic amino acids caused a dramatic reduction of inhibitory activity. Substitutions of hydrophobic amino acids such as Leu or Phe typically caused a significant reduction of the inhibitory activity of synthetic glycoamine analogs. This effect could not be accounted for by the influence of amino acid residue on the biologically active conformation of the carbohydrate constituent of synthetic glycoamine molecules, because the NMR spectra did not reveal any significant difference in a ratio in solution of the major conformational sugar isoforms between biologically active and non-active analogs. Therefore, both carbohydrate and amino acid constituents of the Amadori compounds are essential and necessary in mediating the inhibition of homotypic cancer cell aggregation and clonogenic growth in agarose.

It has been shown recently that integrin-mediated adhesion to the extracellular matrix can rescue cells from apoptosis.[64-67] By analogy, the carbohydrate-mediated adhesion and specific carbohydrate-lectin recognition and binding may activate the anti-apoptosis genes and/or turn off the apoptosis activator genes. It may well be that survival in such a restrictive environment as the dense agarose cultures requires the ability of cells to adhere to the polygalactoside matrix of agarose, and this lectin-mediated adherence may rescue cells from apoptosis. The inhibitors of carbohydrate-lectin recognition and binding would act as antagonists, force cells to go into apoptosis, decrease the fraction of surviving cells and thus cause the reduction of colony-forming efficiency of metastatic cancer cells in a semisolid medium. Recent experimental observations indeed support the notion that the action of cell adhesion inhibitors may force cells into apoptosis,[66-69] and lectin-mediated enhanced homotypic aggregation of metastatic cancer cells may increase their resistance to programmed cell death.[69]

There is growing experimental evidence supporting the concept of anti-cell adhesion therapy for cancer, particularly for treatment of cancer metastases of breast, ovarian, prostate, and colon tumors, as well as malignant melanoma.[14,18-22,70-73] Three prospective approaches are particularly promising: (a) development of synthetic carbohydrate-based competitive adhesion and metastasis inhibitors;[19,63,70,71] (b) biospecific inhibitory monoclonal antibody, lectin-based, and/or peptide-based therapeutics; (c) application of synthetic sugar analogs as oligosaccharide chain modifiers and/or chain terminators.[72] Interference with the carbohydrate-mediated cell-cell recognition and adhesion may lead to reduced growth of primary tumor and decreased metastatic dissemination. The clinical application of anti-cell adhesion therapy may be useful as an adjuvant treatment along with surgery, radiation, and chemotherapy.

References

1. Smets LA. Cell transformation as a model for tumor induction and neoplastic growth. Biochim Biophys Acta 1989; 605:93-111.
2. Von Hoff DD, Clark GM, Stogdill BJ et al. Prospective clinical trial of a human tumor cloning system. Cancer Res 1983; 43:1926-1931.
3. Nomura Y, Tashiro H, Hisamatsu K. Differential effects of estrogen and antiestrogen on in vitro clonogenic growth of human breast cancers in soft agar. J Natl Cancer Inst 1990; 82:1146-1149.
4. Hug V, Rashid R, Johnston D et al. Clonogenic growth and hormone sensitivity of benign and malignant breast tissues. Br J Cancer 1987; 56:619-621.
5. Nomura Y, Tashiro H, Hisamatsu K. In vitro clonogenic growth and metastatic potential of human operable breast cancer. Cancer Res 1989; 49:5288-5293.
6. Cifone MA, Fidler IJ. Correlation of patterns of anchorage-independent growth with in vivo behavior of cells from a murine fibrosarcoma. Proc Natl Acad Sci USA 1980; 77:1039-1043.
7. Price JE. Clonogenicity and experimental metastatic potential of spontaneous mouse mammary neoplasms. J Natl Cancer Inst 1986; 77:529-535.
8. Saiga T, Kashu I, Tabuchi K et al. Increased contraenvironmental pressure division capability: A decisive force in metastasis and invasion of mouse lung adenocarcinoma cell lines. Invasion Metastasis 1989; 9:254-268.
9. Fidler IJ, Li L, Ananthaswamy H et al. Correlation of growth capacity in hard agarose with successful transfection by the activated c-Ha-ras oncogene and in vivo proliferative capacity at metastatic sites. Anticancer Res 1991; 11:17-24.
10. Li L, Price JE, Fan D, Zhang RD, Bucana CD, and Fidler IJ. Correlation of growth capacity of human tumor cells in hard agarose with their in vivo proliferative capacity at specific metastatic sites. J Natl Cancer Inst 1989; 81:1406-1412.
11. Zhang RD, Fidler IJ, Price JE. Relative malignant potential of human breast carcinoma cell lines established from pleural effusions and a brain metastasis. Invasion and Metastasis 1991; 11:204-215.
12. Zhang RD, Price JE, Schackert G, Itoh K, Fidler IJ. Malignant potential of cells isolated from lymph node or brain metastases of melanoma patients and implications for prognosis. Cancer Res 1991; 51:2029-2035.
13. Glinsky GV. Glycoamines: Biochemistry of a new class of humoral tumor markers. J Tumor Markers Oncology 1989; 4:193-221.
14. Glinsky GV. Glycoamines: Structural-functional characterization of a new class of humoral tumor markers. In: Sell S, ed. Serological Cancer Markers. Totowa, NJ: Humana Press, 1992:233-260.
15. Kuo KC, Gehrke JC, Allen WC, Holsbeke M, Li, Zh., Glinsky GV, Zumwalt RW, Gehrke CW. High-performance liquid chromatographic analysis of glycoamines in serum. J Chromatography 1994; 656: 295-302.

16. Glinsky GV, Linetsky MD, Shemyakin VV, Markin VV, Liventsov VV, Ivanova AB. Structural analysis of a new class of humoral tumor markers: Characterization of 48 aminoglycoconjugates (glycoamines) isolated from rat and human blood plasma. J Tumor Marker Oncology 1990; 5:107-118.
17. Glinsky GV, Linetsky MD. Structural analysis of the carbohydrate core of amino-glycoconjugates (glycoamines) from blood plasma of oncological patients. J Tumor Marker Oncology 1990; 5:137-160.
18. Glinsky GV, Linetsky MD, Liventsov VV, Surmilo NI, Ivanova AB, Sidorenko MV, Vinnitsky VB. Glycoamine-synthesizing reaction: Molecular mechanism of the incorporation of monomers into aminoglycoconjugates. J Tumor Marker Oncology 1990; 5:119-136.
19. Glinsky GV. Cell adhesion and metastasis: is the site specificity of cancer metastasis determined by leukocyte-endothelial cell recognition and adhesion? Crit Rev Oncol/Hemat 1993; 14:229-278.
20. Glinsky GV. Glycodeterminants of melanoma cell adhesion. A model for antimetastatic drugs design. Crit Rev Oncol/Hemat 1993; 14:1-13.
21. Glinsky GV. The Blood Group Antigen-Related Glycoepitopes: Biological Functions and Implications in Immunogenesis, Cancer, and AIDS Pathogenesis. Austin, TX: Medical Intelligence Unit, RG Landes Co, 1992.
22. Glinsky GV. The blood group antigens (BGA)-related glycoepitopes. A key structural determinants in immunogenesis and cancer pathogenesis. Crit Rev Oncol/Hemat 1992; 12:151-166.
23. Linetsky MD, Semenova-Kobzar RA, Glinsky GV. The modifying influence of endogenous biopolymers, glycoamines, from the blood serum of BALB/C mice on aggregation properties of experimental rhabdomyosarcoma cells. Experimental Oncology 1991; 13:6, 27-33.
24. Mossine VV, Glinsky GV, Feather MS. The preparation and characterization of some Amadori compounds (1-amino-1-deoxy-D-fructose derivatives) derived from a series of aliphatic ω-amino acids. Carbohydrate Research 1994; 262:257-270.
25. Mossine VV, Glinsky GV, Barnes ChL, Feather MS. Crystal structure of an Amadori compound, N-(1-Deoxy-β-D-fructopyranos-1-yl)-glycine ("D-fructose-glycine"). Carbohydrate Research 1995; 266:5-14.
26. Price JE, Polyzos A, Zhang RD, Daniels LM. Tumorigenicity and metastasis of human breast carcinoma cell lines in nude mice. Cancer Res 1990; 50:717-721.
27. Fidler IJ. The relationship of embolic homogeneity, number, size and viability to the incidence of experimental metastasis. Eur J Cancer 1973; 9:223-227.
28. Fidler IJ. Selection of successive tumor lines for metastasis. Nat New Biol 1973; 242:148-149.
29. Fidler IJ. Biological behavior of malignant melanoma cells correlated to their survival in vivo. Cancer Res 1975; 35:218-224.
30. Hart IR. The selection and characterization of an invasive variant of the B16 melanoma. Am J Pathol 1979; 97:587-592.

31. Meromsky L, Lotan R, Raz A. Implications of endogenous tumor cell surface lectins as mediators of cellular interactions and lung colonization. Cancer Res 1986; 46:5270-5275.
32. Wu V-Y, Cohen P. Receptors specific for Amadori-modified glycated albumin on murine endothelial cells. Biochem Biophys Res Commun 1994; 198:734-739.
33. Nicolson GL. Cancer metastasis. Organ colonization and the cell-surface properties of malignant cells. Biochim Biophys Acta 1982; 695:113-176.
34. Nicolson GL. Cell surface molecules and tumor metastasis. Regulation of metastatic phenotypic diversity. Exp Cell Res 1984; 150:3-22.
35. Roos E. Cellular adhesion, invasion and metastasis. Biochim Biophys Acta 1984; 738:263-284.
36. Glaves D. Correlation between circulating cancer cells and incidence of metastases. Br J Cancer 1983; 48:665-673.
37. Liotta LA, Kleinerman J, Saidel GM. The significance of hematogenous tumor cell clumps in the metastatic process. Cancer Res 1976; 36:889-894.
38. Varani J, Orr W, Ward PA. Adhesive characteristics of tumor cell variant of high and low tumorigenic potential. J Natl Cancer Inst 1980; 64:1173-1178.
39. Winkelhake JL, Nicolson GL. Determination of adhesive properties of variant metastatic melanoma cells to BALB/3T3 cells and their virus-transformed derivatives by a monolayer attachment assays. J Natl Cancer Inst 1976; 56:285-291.
40. Fidler IJ, Bucana C. Mechanism of tumor cell resistance to lysis by syngeneic lymphocytes. Cancer Res 1977; 37:3945-3956.
41. Gasic GL, Gasic TB, Galanti N, Johnson T, Murphy S. Platelet-tumor cell interactions in mice. The role of platelets in the spread of malignant disease. Int J Cancer 1973; 11:704-718.
42. Raz A, Bucana C, McLellan W, Fidler IJ. Distribution of membrane anionic sites on B16 melanoma variants with differing lung colonising potential. Nature (Lond) 1980; 284:363-364.
43. Lotan R, Raz A. Low colony formation in vivo and in culture as exhibited by metastatic melanoma cells selected for reduced homotypic aggregation. Cancer Res 1983; 43:2088-2093.
44. Lotan R, Lotan D, Raz A. Inhibition of tumor cell colony formation in culture by a monoclonal antibody to endogenous lectins. Cancer Res 1985; 45:4349-4353.
45. Lotan R, Cavvalero D, Lotan D, Raz A. Biochemical and immunological characterization of K-1735P melanoma galactoside-binding lectins and their modulation by differentiation inducers. Cancer Res 1989; 49:1261-1268.
46. Raz A, Lotan R. Lectin-like activities associated with human and murine neoplastic cells. Cancer Res 1981; 41:3642-3647.
47. Nicolson GL, Brunson KW, Fidler IJ. Specificity of arrest, survival, and growth of selected metastatic variant cell lines. Cancer Res 1978; 38:4105-4111.

48. Nicolson GL, Winkelhake JL. Organ specificity of blood-borne tumor metastasis determined by cell adhesion? Nature (Lond) 1975; 255:230-232.
49. Allen HJ, Karakousis C, Piver MS, Gamarra M, Nava H, Forsyth B, Matecki B, Jazayeri A, Sucato D, Kisailus E, DiCioccio R. Galactoside-binding lectin in human tissues. Tumor Biol 1987; 8:218-229.
50. Carding SR, Thorpe SJ, Thorpe R, Feizi T. Transformation and growth related changes in levels of nuclear and cytoplasmic proteins antigenically related to mammalian β-galactoside-binding lectin. Biochem Biophys Res Commun 1985; 127:680-686.
51. Gabius H-J, Engelhardt R, Cramer F. Endogenous tumor lectins: Overview and perspectives. Anticancer Res 1986; 6:573-578.
52. Gitt MA, Barondes SH. Evidence that a human soluble β-galactoside-binding lectin is encoded by a family of genes. Proc Natl Acad Sci USA 1986; 83:7603-7607.
53. Raz A, Meromsky L, Carmi P, Karakash R, Lotan D, Lotan R. Monoclonal antibodies to endogenous galactoside-specific tumor cell lectins. EMBO J 1984; 3:2979-2983.
54. Raz A, Meromsky L, Lotan R. Differential expression of endogenous lectins on the surface of nontumorigenic, tumorigenic, and metastatic cells. Cancer Res 1986; 46:3667-3672.
55. Raz A, Avivi A, Pazarini G, Garmi P. Cloning and expression of cDNA for two endogenous UV-2237 fibrosarcoma lectin genes. Exp Cell Res 1987; 173:109-116.
56. Raz A, Meromsky L, Zvibel I, Lotan R. Transformation-related changes in the expression of endogenous cell lectins. Int J Cancer 1987; 39:353-360.
57. Raz A, Carmi P, Pazarini G. Expression of two different endogenous galactoside-binding lectins sharing sequence homology. Cancer Res 1988; 48:645-649.
58. Weis WI. Recognition of cell surface carbohydrates by C-type animal lectins. In: Metcalf BW, Dalton BJ, Poste G, eds. Cellular Adhesion: Molecular Definition to Therapeutic Potential. New York: Plenum Press, 1994:55-75.
59. Springer GF. T and Tn, general carcinoma autoantigens. Science 1984; 224:1198-1206.
60. Stein R, Chen S, Grossman W, Goldenberg DM. Human lung carcinoma monoclonal antibody specific for the Thomsen-Friedenreich antigen. Cancer Res 1989; 49:32-37.
61. Gluzman DF, Abramenko IB, Sklyarenko LM, Pisnyachevskaya GV. Immunocytochemical Diagnosis of The Malignant Serous Effusions. Kiev: Naukova Dumka, 1993.
62. Springer GF, Cheinsong-Popov R, Schirrmacher V, Desoi PR, Tegtmeyer H. Proposed molecular basis of murine tumor cell-hepatocyte interaction. J Biol Chem 1983; 258:5702-5706.
63. Lee RT, Lin P, Lee YuCh. New synthetic cluster ligands for galactose/N-acetylgalactosamine-specific lectin of mammalian liver. Biochemistry 1984; 23:4225-4261.

64. Ruoslahti E, Reed J. Anchorage dependence, integrins, and apoptosis. Cell 1994; 77:477-478.
65. Boudreau N, Sympson CJ, Werb Z, Bissell MJ. Suppression of ICE and apoptosis in mammary epithelial cells by extracellular matrix. Science 1995; 267:891-893.
66. Frisch SM, Francis H. Disruption of epithelial cell-matrix interactions induces apoptosis. J Cell Biol 1994; 124:619-626.
67. Meredith JE, Fazeli B, Schwartz MA. The extracellular matrix as a cell survival factor. Mol Biol Cell 1994; 4:953-961.
68. Brooks PC, Montgomery AMP, Rosenfeld M, Reisfeld RA, Hu T, Klier G, Cheresh DA. Integrin $\alpha_v\beta_3$ antagonists promote tumor regression by inducing apoptosis of angiogenic blood vessels. Cell 1994; 79:1157-1164.
69. Glinsky GV, Glinsky VV. Apoptosis and metastasis: A superior resistance of metastatic cancer cells to programmed cell death. Cancer Lett 1996; 101:43-51.
70. Hakomori S-I. Possible functions of tumor-associated carbohydrate antigens. Curr Opin Immunol 1991; 3:646-653.
71. Hakomori S-I. Possible new directions in cancer therapy based on aberrant expression of glycosphingolipids: Anti-adhesion and ortho-signaling therapy. Cancer Cells 1991; 3(12):461-470.
72. Woynarowska B, Skrincosky DM, Haag A, Sharma M, Matta K, Bernacki RJ. Inhibition of lectin-mediated ovarian tumor cell adhesion by sugar analogs. J Biol Chem 1994; 269:22797-22803.
73. Pienta KJ, Naik H, Akhtar A et al. Inhibition of spontaneous metastasis in a rat prostate cancer model by oral administration of modified citrus pectin. J Natl Cancer Inst 1995; 87:348-353.
74. Glinsky GV, Mossine VV, Price JE, Bielenberg D, Glinsky VV, Ananthaswamy HN, Feather MS. Inhibition of colony formation in agarose of metastatic human breast carcinoma and melanoma cells by synthetic glycoamine analogs. Clin Exper Metastasis 1996; 14:253-267.

CHAPTER 6

Inhibition of Human Breast Cancer Metastasis in Nude Mice by Synthetic Glycoamines

6.1. Summary

In this chapter we describe experiments in which we directly tested the concept of anti-cell adhesion therapy of cancer metastasis. We employed the most advanced preclinical model available—the nude mouse model of human breast cancer metastasis, and utilized the highly metastatic human breast cancer cell line MDA-MB-435. We examined the effect of synthetic low molecular weight glycoamine analogs on the metastasis of MDA-MB-435 human breast carcinoma xenografts growing in the mammary fat pads (m.f.p.) of nude mice. Initial in vitro screening of a panel of synthetic glycoamines was performed employing a clonogenic growth assay in 0.9% agarose. Eight of nine compounds manifested a significant dose-dependent inhibition of colony formation by MDA-MB-435 cells in 0.9% agarose. The relative activity ranking of the compounds, based on 50% inhibition (I_{50}) doses independently determined for each synthetic glycoamine analogs, identified Lac-L-Leu, Fru-D-Leu, Fru-L-Phe, and Fru-L-Leu as the most effective inhibitors of colony formation. Two separate experimental treatment protocols were used to examine the effect of selected synthetic glycoamines on human breast cancer growth and metastasis in athymic nude mice. Group A mice were treated intraperitoneally (i.p.) daily from day 2 after injection of the breast cancer cells, until the end of the experiment (17 weeks). In group B, the mice were untreated until the mean tumor diameter was 10 mm, at which time daily i.p. treatment began. After 7 days, the primary tumors were resected, and the mice treated for a further 4

Failure of Apoptosis and Cancer Metastasis, by Gennadi V. Glinsky.

weeks (a total of 5 weeks of treatment). The synthetic glycoamines did not have a significant anti-tumor effect, and there was no difference in the tumor incidence or tumor growth rates in mice treated continuously with either synthetic glycoamines or PBS. The significant antimetastatic activity of synthetic glycoamines was detected in both experimental treatment protocols. In mice continuously treated with synthetic glycoamines according to protocol A, the incidence of metastasis decreased 4.6-fold ($p=0.014$) and 2.7-fold ($p=0.031$) in mice treated with Fru-D-Leu and Lac-L-Leu, respectively. In mice in protocol B, the incidence of pulmonary metastasis was decreased 1.9-fold ($p=0.069$) and 2.5-fold ($p=0.042$) in mice treated with Fru-D-Leu and Lac-L-Leu, respectively. Correspondingly, the average number of spontaneous pulmonary metastases was reduced from 37 in control mice to 0.2 ($p=0.005$) and 0.9 ($p<0.02$) in mice treated according to protocol A with Fru-D-Leu and Lac-L-Leu, respectively. Treatment of mice with Fru-L-Leu did not have a significant anti-metastatic effect, and no reduction in metastasis incidence or number was noted in mice treated with this synthetic glycoamine analog. The treated animals had no apparent toxicity from chronic daily injection (up to 17 weeks of treatment) of synthetic glycoamines, and no obvious pathology was noted in the histological slides of the livers, kidneys or spleens of the treated mice. Therefore, we identified two synthetic glycoamines (Fru-D-Leu and Lac-L-Leu) that were effective inhibitors of spontaneous human breast cancer metastasis in nude mice. Potential mechanisms for antimetastatic activity of synthetic glycoamines may include the inhibition of β-galectin-mediated homotypic cancer cell aggregation and induction of apoptosis in target cells.

6.2. Introduction

Breast cancer is the most common cancer among women in North America and western Europe and is the second leading cause of female cancer deaths in the United States.[1,2] In the U.S., age-adjusted breast cancer incidence rates increased considerably during this century.[3] Approximately one in every nine American women will develop breast cancer during her lifetime, as breast cancer incidence rates in the U.S. have increased about 3% a year since 1980. An estimated 180,000+ new breast cancer cases were diagnosed in the U.S. in 1995 and over 46,000 deaths from breast cancer were reported.[2] Approximately 40% of patients diagnosed with breast cancer have disease that has regional or distant metastases and, at present, there is no efficient curative therapy for breast cancer patients with advanced metastatic disease.

The anchorage-independent growth of primary breast cancer cells in a semisolid medium was found to be an independent variable associated with shorter disease-free survival and overall survival after mastectomy.[4,5] The clonogenic growth in agar cultures was also reported to correlate with the progression of breast neoplasms from benign to malignant.[4,5] In several experimental systems using rodent cells, colony forming efficiency in agarose was shown to be correlated with metastatic potential,[6,7] and a particularly discriminating assay was to test colony formation in higher densities of agar medium.[8,9] It has also been reported that for human tumor cell lines, including breast cancer cells, the ability to grow in dense agarose cultures (>0.6% agarose) corresponds well with metastatic potential tested in nude mice.[9-12] Thus, the colony formation assay in agarose cultures of greater than 0.6% density distinguishes cancer cells with high metastatic potential.[10-12] Therefore, the clonogenic growth assay could be used for in vitro screening of potential antimetastatic compounds.

The glycoamines are naturally occurring low molecular weight components of mammalian blood serum that were studied as potential humoral cancer markers.[13-15] The prototype compounds were originally isolated from human and rodent blood serum, employing a combination of conventional chromatography, analytical ultrafiltration, and size-exclusion and RP HPLC.[13-18] Subsequently, the compounds were structurally characterized as glycoconjugates of amino acids with covalent structures of Schiff base, ester, and Amadori bonding.[13,15-18] The inhibitory activity of HPLC-purified natural and synthetic glycoamines on rodent cancer cell aggregation in vitro and experimental metastasis in vivo was reported.[13,14,19-22] Recently, we described a method of synthesis, purification and full structural characterization of a series of synthetic glycoamine analogs, using NMR, MS and X-ray analysis.[23-25] The availability of synthetic glycoamines in pure form in multigram quantities allowed us to demonstrate a high inhibitory effect of the synthetic glycoamines on clonogenic growth in agarose of metastatic human cancer cells.[25] The main objective of this study was to select the most effective synthetic glycoamines acting as inhibitors of clonogenic growth of metastatic breast carcinoma cells in hard agarose and to evaluate the antimetastatic activity of selected glycoamine analogs using the nude mouse model of human breast cancer metastasis.[11,26]

We here report significant inhibition (over 70%) of the incidence of spontaneous human breast cancer metastasis in the nude mouse model by synthetic glycoamine analogs. We also report that antimetastatic synthetic glycoamines inhibit homotypic aggregation of human MDA-MB-435 breast carcinoma cells and induce apoptosis in metastatic cancer cells.

6.3. Selection of Inhibitors of Colony-Formation by MDA-MB-435 Cells in Dense Agarose

We previously reported the inhibition of colony formation by MDA-MB-435 cells following exposure to a panel of nine synthetic glycoamines.[25] Figure 6.1 summarizes the results that lead to the identification of four compounds that were most effective in the in vitro assay; Lac-L-Leu, Fru-D-Leu, Fru-L-Phe and Fru-L-Leu. The three Leu analogs were selected for the in vivo study. The complete analysis of in vitro selection protocol is provided in the previous chapter.

6.4. Nude Mouse-Human Breast Cancer Xenograft Experiments

Two experimental treatment protocols were used to determine the in vivo activity of the three selected synthetic glycoamines. In protocol A, the mice were treated daily from 2 days after tumor cell injection until the end of the experiment (17 weeks). In group B, daily treatment started when the mean tumor diameter was 10 mm. After 7 days of treatment, the primary tumors were removed, and the mice treated for a further 4 weeks (a total of 5 weeks of treatment).

Effect of Synthetic Glycoamines on Incidence and Growth of Primary Tumors

The growth curves of MDA-MB-435 mfp tumors in mice treated continuously from day 2 after cancer cell injection with Fru-D-Leu or Fru-L-Leu did not differ from the growth curves of the PBS-treated mice (data not shown). The weights of tumors for the control and treated mice at time of autopsy were similar and the mean weights showed no statistical difference. The incidence of tumors in these mice did not differ, with 82% incidence in the PBS-treated group, and 75% incidence in mice treated with either Fru-D-Leu or Fru-L-

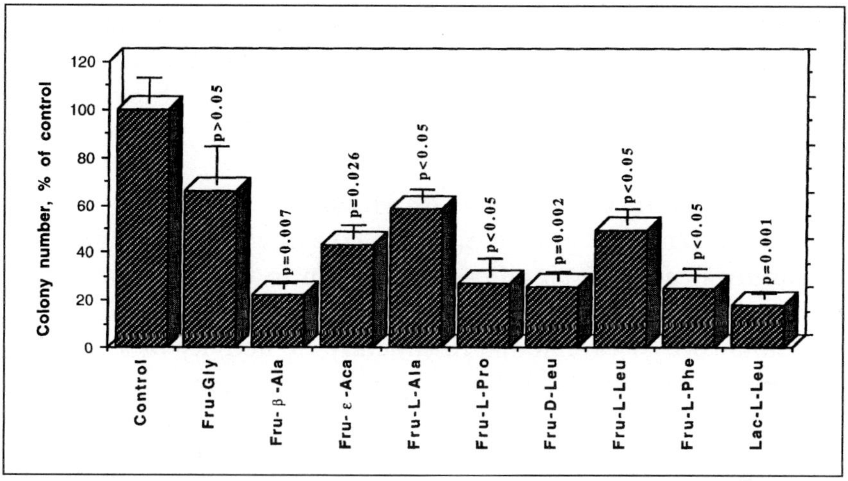

Fig. 6.1. Effect of synthetic glycoamines on MDA-MB-435 colony formation in 0.9% agarose. Clonogenic growth assay in agarose was performed as described.[25] The presence of synthetic glycoamines significantly inhibited the colony formation ability in hard agarose of MDA-MB-435 cells as compared with control PBS-treated cells. The results are presented as the mean and SE of triplicate experiments. Reproduced with permission from Glinsky GV et al, Cancer Res 1996; 56:5319-5324.

Leu. In mice that were scored as not having tumor at the end of the experiment, either no tumor was found at the site of injection, or a mass of 2-3 mm that had shown no progressive growth. Histological evaluation of the small masses showed residual tumor cells, surrounded by fibrous connective tissues. The growth curves of MDA-MB-435 mfp tumors in mice treated continuously with Lac-L-Leu showed a trend toward inhibition of the local tumor, although a comparison of the weights of tumors removed from these mice at the time of autopsy (control mean weight = 0.69 ± 0.6g (Mean ± SD), test mean weight = 0.44 ± 0.54g) showed no statistically significant difference (Student's t-test). A 100% incidence of tumors was recorded in group A mice treated with Lac-L-Leu.

Thus, the synthetic glycoamines did not have a significant antitumor effect, and there was no difference in the tumor incidence or tumor growth rates of MDA-MB-435 tumors in mice treated continuously with synthetic glycoamines or with PBS.

Effect of Synthetic Glycoamines on Tumor Host

The treated animals had no apparent toxicity from the daily injection (up to 17 weeks of treatment) of synthetic glycoamines. Animal weights, skin texture, and overall behavior were unchanged between the control and the treated groups. No obvious pathology was noted in the histological slides of the livers, kidneys or spleens of the treated mice. Thus, chronic treatment with synthetic glycoamines did not apparently lead to toxicity.

Effect of Synthetic Glycoamines on Metastasis

As shown in Figure 6.2, daily treatment with synthetic glycoamines caused a significant reduction in both the incidence and number of spontaneous pulmonary metastases of MDA-MB-435 human breast carcinoma. Antimetastatic activity of synthetic glycoamines was detected in both experimental treatment protocols. Only 18% to 22% of control mice (4 of 18 control mice in group A and 3 of 17 control mice in group B) failed to develop pulmonary metastases, compared with 83% to 57% of mice treated with Fru-D-Leu (5 of 6 treated mice in group A and 4 of 7 treated mice in group B) or 71% to 67% of mice treated with Lac-L-Leu (5 of 7 treated mice in group A and 4 of 6 treated mice in group B). In mice continuously treated with synthetic glycoamines according to protocol A, the incidence of metastasis was decreased 4.6-fold ($p=0.014$) and 2.7-fold ($p=0.031$) in mice treated with Fru-D-Leu and Lac-L-Leu, respectively (Fig. 6.2). The antimetastatic activity of synthetic glycoamines was less profound in mice treated according to protocol B, probably due to the significantly shorter treatment period (a total of 17 and 5 weeks of treatment in protocol A and B, respectively). In mice treated daily with synthetic glycoamines according to protocol B, the incidence of pulmonary metastasis was decreased 1.9-fold ($p=0.069$) and 2.5-fold ($p=0.042$) in mice treated with Fru-D-Leu and Lac-L-Leu, respectively (Fig. 6.2). Correspondingly, the average number of spontaneous pulmonary metastases was reduced from 37 in control mice to 0.2 ($p=0.005$) and 0.9 ($p<0.02$) in mice treated according to protocol A with Fru-D-Leu and Lac-L-Leu, respectively.[35] A similar trend toward inhibition of the number of pulmonary metastasis in mice treated according to protocol B was noticed in mice treated with Fru-D-Leu and Lac-L-Leu.[35] Treatment of mice with Fru-L-Leu did not have a significant anti-metastatic effect, and no reduction in metastasis incidence or number were noted in mice treated with this synthetic glycoamine analog.

Therefore, we have identified two synthetic glycoamines (Fru-D-Leu and Lac-L-Leu) that were effective inhibitors of spontaneous human breast cancer metastasis in nude mice.

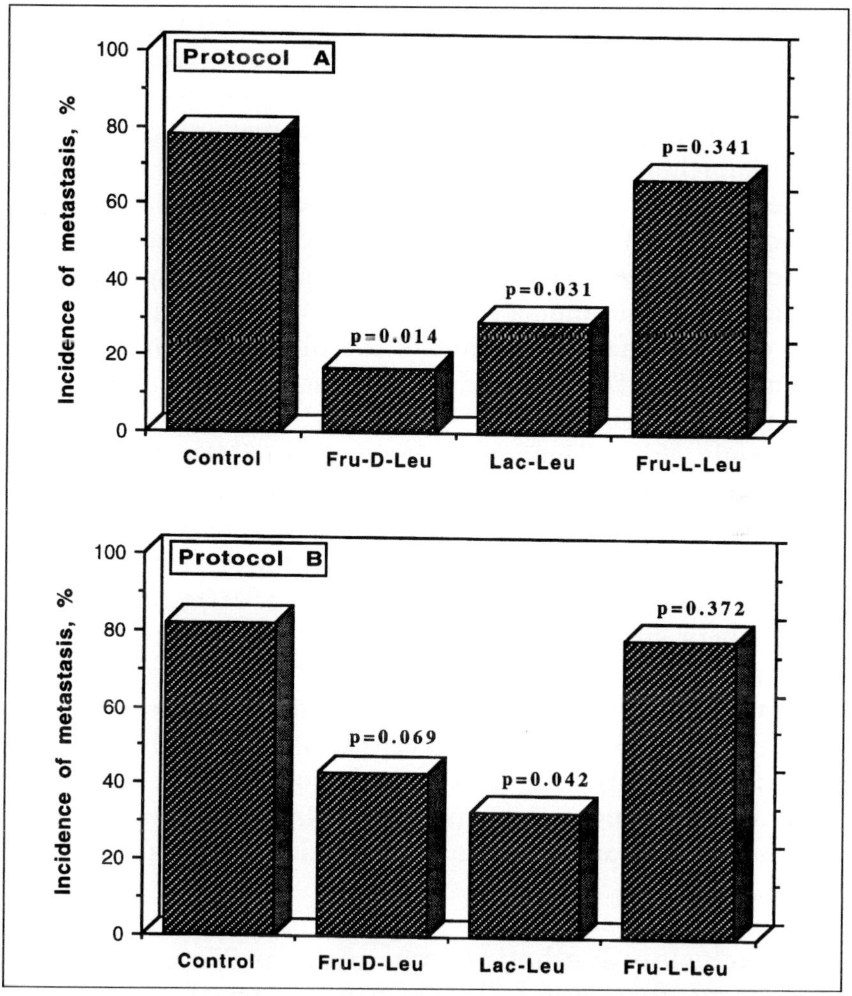

Fig. 6.2. Inhibition of MDA-MB-435 human breast cancer metastasis in nude mice by synthetic glycoamines. MDA-MB-435 cells were injected into the mfp of nude mice on day 0 (per institutionally approved protocol). Animals were given daily i.p. injections of synthetic glycoamines or PBS (control) according to the treatment protocols A and B;[1] after the treatment the animals were killed and necropsies were performed.
A. Number of mice with spontaneous lung metastases was significantly reduced in the FDL- and LL-treated groups in accordance with protocol A. The mice were treated i.p. daily from day 2 after injection of MDA-MB-435 cells, until the end of the experiment (17 weeks).
B. Number of mice with spontaneous lung metastases was significantly reduced in the FDL- and LL-treated groups in accordance with protocol B. The mice were untreated until the mean tumor diameter was 10 mm, at which time daily i.p. treatment began. After 7 days, the tumors were resected, and the mice were treated for a further four weeks (total of five weeks of treatment).
P values were calculated using the Fisher exact test.

6.5. Inhibition of Homotypic Aggregation and Induction of Apoptosis in MDA-MB-435 Human Breast Carcinoma Cells by Synthetic Glycoamines

It had been previously demonstrated that synthetic glycoamines could inhibit homotypic aggregation of murine B16 melanoma cells by interfering with β-galactoside-specific lectin-mediated cell-cell interactions.[25] Blocking of cell surface galectin-mediated cell-cell interactions resulted in induction of apoptosis in a highly metastatic B16-F10 murine melanoma cell variant.[32] We investigated whether these mechanisms are relevant to the bioactivity of synthetic glycoamines on MDA-MB-435 human breast carcinoma cells as well. Similar to the murine B16 melanoma model, synthetic glycoamines significantly inhibited homotypic aggregation of MDA-MB-435 cells and blocked binding of β-galactoside-specific PNA lectin to the MDA-MB-435 cells (Fig. 6.3). Over 50% reduction in homotypic aggregation of human breast carcinoma cells treated with 5 mM of Fru-D-Leu and Lac-L-Leu was observed. Treatment of MDA-MB-435 cells with synthetic glycoamines caused a four- to six-fold increase in the apoptosis level as determined by the TUNEL assay (Fig. 6.4A). Induction of apoptosis in glycoamine-treated highly metastatic MDA-MB-435 human breast carcinoma cells was confirmed using DNA fragmentation analysis, which clearly demonstrated an accumulation of low molecular weight apoptotic DNA in MDA-MB-435 cells at 48 hours after treatment with antimetastatic synthetic glycoamines (Fig. 6.4B).

6.6. Conclusion: Synthetic Glycoamines as Prototype Compounds for a Novel Class of Antimetastatic Drugs

We have identified two synthetic glycoamines that were effective inhibitors of spontaneous human breast cancer metastasis in nude mice. Continuous intraperitoneal injection of Lac-L-Leu or Fru-D-Leu, beginning 2 days after the inoculation of MDA-MB-435 human breast carcinoma cells in mammary fat pads of athymic nude mice, decreased the incidence of pulmonary metastasis by 2.7- and 4.6-fold, respectively. The average number of spontaneous pulmonary metastases was reduced from 37 in PBS-treated mice to 0.9 in Lac-L-Leu-treated mice and 0.2 in Fru-D-Leu-treated mice. The relative number of metastasis-free mice (without either macroscopic and microscopic metastases) was increased 3.8-fold in Fru-D-Leu-treated mice and 3.2-fold in Lac-L-Leu-treated mice. Chronic treatment (daily i.p. injection during 5 or 17 weeks of treatment) with synthetic glycoamines did not lead to any apparent toxicity.

In vitro screening of potential antimetastatic compounds using colony formation assay in dense agarose could be useful for selecting promising compounds for subsequent in vivo studies. According to our results the most promising compounds were those that produced over 70%-80% inhibition of the clonogenic growth of the MDA-MB-435 cells in dense agarose. The effective in vivo antimetastatic compounds, Fru-D-Leu and Lac-L-Leu, caused 74% and 82% inhibition, respectively, of colony formation in 0.9% agarose. Fru-L-Leu showed a 51% inhibition of colony formation in 0.9% agarose by MDA-MB-435 cells, yet did not produce any significant anti-metastatic effect in vivo.

The synthetic glycoamines did not have a significant effect on tumor growth and there was no difference in the primary tumor incidence of MDA-MB-435 mfp tumors in treated or control group mice. This observation is not surprising because synthetic glycoamines were described as inhibitors of cancer cell aggregation and adhesion mediated by cell surface β-galactoside-specific lectins.[19,25] Carbohydrate-mediated homotypic and heterotypic cancer cell aggregation and adhesion is thought to be one of the key eventS in the metastatic process,[19-22,23] which apparently does not play a significant role in growth of primary neoplasms. A similar selective antimetastatic effect without noticeable anti-tumor activity has been recently reported for another carbohydrate-based metastasis inhibitor. A modified citrus pectin was reported to act by interfering with β-galectin-mediated adhesion of metastatic cancer cells in a rat prostate carcinoma spontaneous metastasis model.[34]

Two apparently unrelated chemical modifications of the Fru-L-Leu molecule (substitution of L-Leu for D-Leu in Fru-D-Leu and addition of a βGal residue in Lac-L-Leu) resulted in a dramatic increase in biological activities, as determined in both the clonogenic growth assay and the nude mouse experiments. It will be interesting to examine if a combination of these two chemical modifications would lead to the synthesis of even more effective antimetastatic compounds. The observation that synthetic glycoamines induce apoptosis in vitro in human metastatic breast cancer cells certainly may add a new dimension to our understanding of the molecular

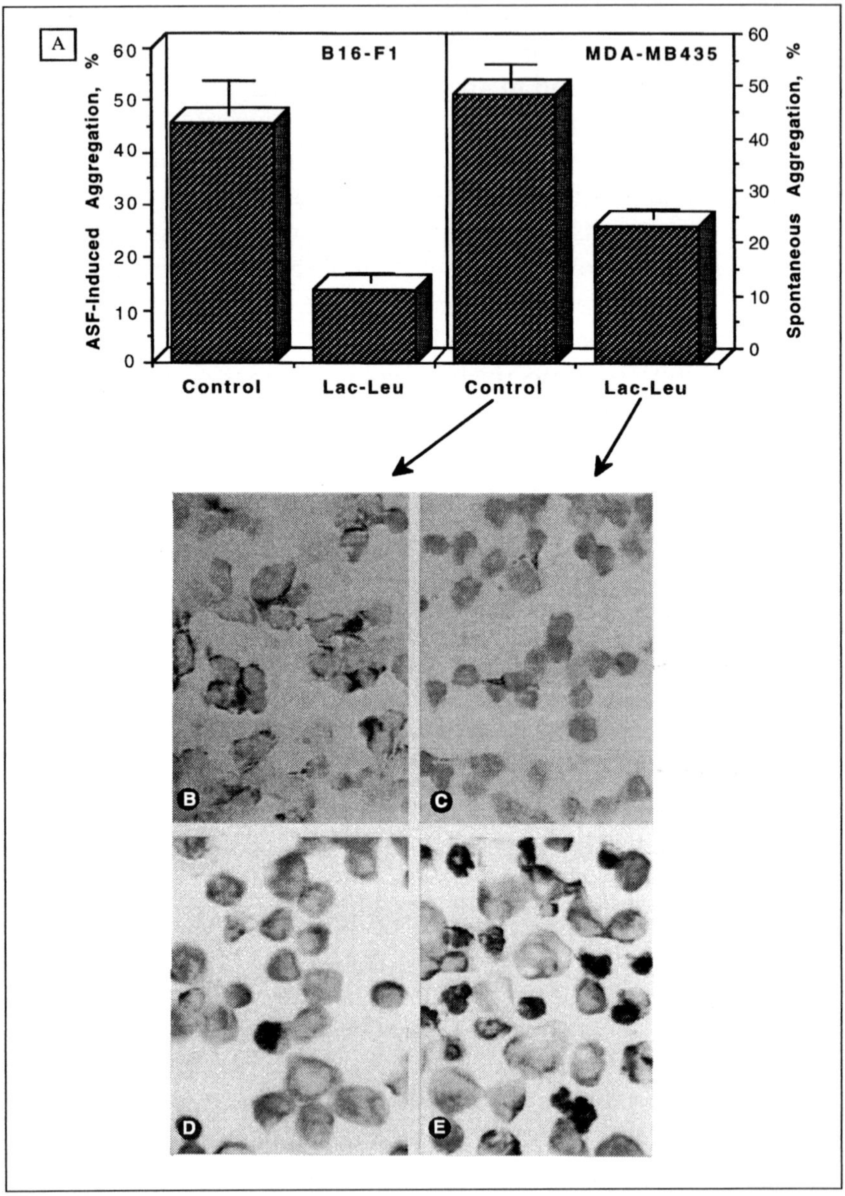
A
B16-F1
MDA-MB435
ASF-Induced Aggregation, %
Spontaneous Aggregation, %
0
10
20
30
40
50
60
Control
Lac-Leu
Control
Lac-Leu
B
C
D
E

Fig. 6.3 (opposite). Inhibition of ASF-induced B16-F1 murine melanoma and spontaneous MDA-MB-435 human breast carcinoma cell aggregation (A) and binding of β-galactoside-specific PNA lectin to the MDA-MB-435 cells (B and C) is accompanied by induction of apoptosis in target cells (D and E). Cell aggregation (A), lectin binding (B and C), and apoptosis induction (D and E) experiments were performed as described.[25,35] The evaluation of both cell aggregation and PNA lectin-binding activities was performed after one hour of incubation of MDA-MB-435 cells. The apoptosis level was determined employing the TUNEL assay after two days of cultivation of MDA-MB-435 cells in RPMI 1640 medium containing 0.1% FCS without (control) and with addition of 2 mM (final concentration) of synthetic glycoamines. Representative results of one of two similar experiments are presented. Reproduced with permission from Glinsky GV et al, Cancer Res 1996; 56:5319-5324.

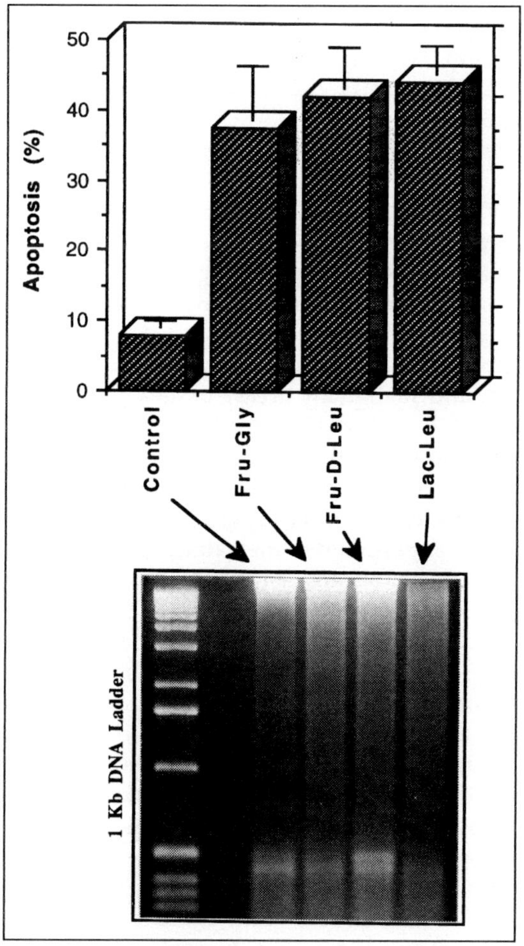

Fig. 6.4. Induction of apoptosis in MDA-MB-435 human breast carcinoma cells by antimetastatic synthetic glycoamines. MDA-MB-435 cells were incubated in RPMI 1640 medium containing 0.1% FCS without (control) and with addition of 2 mM (final concentration) synthetic glycoamines. After 2 days of cultivation the apoptosis level was determined employing the TUNEL assay (A) and DNA fragmentation analysis by electrophoresis in 1.2% agarose gel and ethidium bromide staining (B) as described.[35] The counts of apoptotic cells were made on a minimum of 200 cells per culture. Representative results from one of two similar experiments are presented as a mean and SD of triplicate determinations. Reproduced with permission from Glinsky GV et al, Cancer Res 1996; 56:5319-5324.

mechanisms of the biological activity of this class of molecules. However, it is not yet clear how the in vitro apoptosis induction activity of synthetic glycoamines is related to their in vivo antimetastatic activity and, obviously, further experiments are needed along this line.

Our data further support the concept of anti-cell adhesion therapy for cancer, particularly for treatment of cancer metastases of breast, ovarian, prostate, and colon tumors and malignant melanoma.[19-22,25,33,34] Interference with carbohydrate-mediated cell-cell recognition and adhesion may lead to reduced metastatic dissemination and may be useful as an adjuvant treatment along with surgery, radiation, and chemotherapy.

References

1. Howe GR, Hirohata T, Hislop TG et al. Dietary factors and risk of breast cancer: combined analysis of 12 case-control studies. J Natl Cancer Inst 1990; 82:561-569.
2. Wingo PA, Tong T, Bolden S. Cancer Statistics 1995. CA Cancer J Clin 1995; 45:8-30.
3. Harris JR, Lippman ME, Veronesi U et al. Breast cancer (part 1). N Engl J Med 1992; 327:319-328.
4. Nomura Y, Tashiro H, Hisamatsu K. In vitro clonogenic growth and metastatic potential of human operable breast cancer. Cancer Res 1989; 49:5288-5293.
5. Hug V, Rashid R, Johnston D et al. Clonogenic growth and hormone sensitivity of benign and malignant breast tissues. Br J Cancer 1987; 56:619-621.
6. Cifone MA, Fidler IJ. Correlation of patterns of anchorage-independent growth with in vivo behavior of cells from a murine fibrosarcoma. Proc Natl Acad Sci USA 1980; 77:1039-1043.
7. Price JE. Clonogenicity and experimental metastatic potential of spontaneous mouse mammary neoplasms. J Natl Cancer Inst 1986; 77:529-535.
8. Saiga T, Kashu I, Tabuchi K et al. Increased contraenvironmental pressure division capability: A decisive force in metastasis and invasion of mouse lung adenocarcinoma cell lines. Invasion & Metastasis 1989; 9:254-268.
9. Fidler IJ, Li L, Ananthaswamy H et al. Correlation of growth capacity in hard agarose with successful transfection by the activated c-Ha-ras oncogene and in vivo proliferative capacity at metastatic sites. Anticancer Res 1991; 11:17-24.
10. Li L, Price JE, Fan D, Zhang RD, Bucana CD, Fidler IJ. Correlation of growth capacity of human tumor cells in hard agarose with their in vivo proliferative capacity at specific metastatic sites. J Natl Cancer Inst 1989; 81:1406-1412.

11. Zhang RD, Fidler IJ, Price JE. Relative malignant potential of human breast carcinoma cell lines established from pleural effusions and a brain metastasis. Invasion & Metastasis 1991; 11:204-215.
12. Zhang RD, Price JE, Schackert G, Itoh K, Fidler IJ. Malignant potential of cells isolated from lymph node or brain metastases of melanoma patients and implications for prognosis. Cancer Res 1991; 51:2029-2035.
13. Glinsky GV. Glycoamines: Biochemistry of a new class of humoral tumor markers. J Tumor Markers Oncology 1989; 4:193-221.
14. Glinsky GV. Glycoamines: Structural-functional characterization of a new class of humoral tumor markers. In: Serological Cancer Markers. Sell S, ed. Totowa, NJ: Humana Press, 1992:233-260.
15. Kuo KC, Gehrke JC, Allen WC, Holsbeke M, Li, Zh., Glinsky GV, Zumwalt RW, Gehrke CW. High-performance liquid chromatographic analysis of glycoamines in serum. J Chromatography 1994; 656:295-302.
16. Glinsky GV, Linetsky MD, Shemyakin VV, Markin VV, Liventsov VV, Ivanova AB. Structural analysis of a new class of humoral tumor markers: Characterization of 48 aminoglycoconjugates (glycoamines) isolated from rat and human blood plasma. J Tumor Marker Oncology 1990; 5:107-118.
17. Glinsky GV, Linetsky MD. Structural analysis of the carbohydrate core of amino-glycoconjugates (glycoamines) from blood plasma of oncological patients. J Tumor Marker Oncology 1990; 5:137-160.
18. Glinsky GV, Linetsky MD, Liventsov VV, Surmilo NI, Ivanova AB, Sidorenko MV, Vinnitsky VB. Glycoamine-synthesizing reaction: Molecular mechanism of the incorporation of monomers into aminoglycoconjugates. J Tumor Marker Oncology 1990; 5:119-136.
19. Glinsky GV. Cell adhesion and metastasis: is the site specificity of cancer metastasis determined by leukocyte-endothelial cell recognition and adhesion? Crit Rev Oncol/Hemat 1993; 14:229-278.
20. Glinsky GV. Glycodeterminants of melanoma cell adhesion. A model for antimetastatic drugs design. Crit Rev Oncol/Hemat 1993; 14:1-13.
21. Glinsky GV. The Blood Group Antigen-Related Glycoepitopes: Biological Functions and Implications in Immunogenesis, Cancer, and AIDS Pathogenesis. Austin, TX: Medical Intelligence Units, RG Landes Co., 1992.
22. Glinsky GV. The blood group antigens (BGA)-related glycoepitopes. A key structural determinants in immunogenesis and cancer pathogenesis. Crit Rev Oncol/Hemat 1992; 12:151-166.
23. Mossine VV, Glinsky GV, Feather MS. The preparation and characterization of some Amadori compounds (1-amino-1-deoxy-D-fructose derivatives) derived from a series of aliphatic ω-amino acids. Carbohydrate Research 1994; 262:257-270.
24. Mossine VV, Glinsky GV, Barnes ChL, Feather MS. Crystal structure of an Amadori compound, N-(1-Deoxy-β-D-fructopyranos-1-yl)-glycine ("D-fructose-glycine"). Carbohydrate Research 1995; 266:5-14.

25. Glinsky GV, Mossine VV, Price JE, Bielenberg D, Glinsky VV, Ananthaswamy HN, Feather MS. Inhibition of colony formation in agarose of metastatic human breast carcinoma and melanoma cells by synthetic glycoamine analogs. Clin & Exper Metastasis 1996; 14:253-267.
26. Price JE, Polyzos A, Zhang RD, Daniels LM. Tumorigenicity and metastasis of human breast carcinoma cell lines in nude mice. Cancer Res 1990; 50:717-721.
27. Fidler IJ. The relationship of embolic homogeneity, number, size and viability to the incidence of experimental metastasis. Eur J Cancer 1973; 9:223-227.
28. Fidler IJ. Selection of successive tumor lines for metastasis. Nat New Biol 1973; 242:148-149.
29. Fidler IJ. Biological behavior of malignant melanoma cells correlated to their survival in vivo. Cancer Res 1975; 35:218-224.
30. Hart IR. The selection and characterization of an invasive variant of the B16 melanoma. Am J Pathol 1979; 97:587-592.
31. Zar JH. Biostatistical Analysis, 2nd ed. Englewood Cliffs, NJ: Prentice-Hall, Inc., 1984.
32. Glinsky GV, Glinsky VV. Apoptosis and metastasis: A superior resistance of metastatic cancer cells to programmed cell death. Cancer Letters 1996; 101:43-51.
33. Meromsky L, Lotan R, Raz A. Implications of endogenous tumor cell surface lectins as mediators of cellular interactions and lung colonization. Cancer Res 1986; 46:5270-5275.
34. Pienta KJ, Naik H, Akhtar A et al. Inhibition of spontaneous metastasis in a rat prostate cancer model by oral administration of modified citrus pectin. J Natl Cancer Inst 1995; 87:348-353.
35. Glinsky GV, Price JE, Glinsky VV, Mossine VV, Kiriakova G, Metcalf JB. Inhibition of human breast cancer metastasis in nude mice by synthetic glycoamines. Cancer Res 1996; 56:5319-5324.

Index

L

M

N

O

P

R

S

T

V

Z